Recent Advances in Thin Film Electronic Devices

Recent Advances in Thin Film Electronic Devices

Editor

Chengyuan Dong

MDPI • Basel • Beijing • Wuhan • Barcelona • Belgrade • Manchester • Tokyo • Cluj • Tianjin

Editor
Chengyuan Dong
Department of Electronic
Engineering
Shanghai Jiao Tong
University
Shanghai
China

Editorial Office
MDPI
St. Alban-Anlage 66
4052 Basel, Switzerland

This is a reprint of articles from the Special Issue published online in the open access journal *Micromachines* (ISSN 2072-666X) (available at: www.mdpi.com/journal/micromachines/special_issues/Thin_Film_Electronic).

For citation purposes, cite each article independently as indicated on the article page online and as indicated below:

LastName, A.A.; LastName, B.B.; LastName, C.C. Article Title. *Journal Name* **Year**, *Volume Number*, Page Range.

ISBN 978-3-0365-5294-1 (Hbk)
ISBN 978-3-0365-5293-4 (PDF)

Contents

About the Editor

Chengyuan Dong

Chengyuan Dong, associate professor in Department of Electronic Engineering, Shanghai Jiao Tong University. He obtained a Ph.D. degree from Shanghai Jiao Tong University in 2003, and then served some FPD makers in China mainland until 2008. His current research interest is macroelectronics, including physics and fabrication of novel TFT devices, as well as design and application of macroelectronic circuits. Dr. Dong is a committee member in SID Beijing-Chapter and a Guest Editor for *Micromachines*.

Preface to "Recent Advances in Thin Film Electronic Devices"

Thin film electronic devices have been becoming a hot topic with the rapid development of the related industries, such as flat panel display, flat panel sensors, energy devices, memories, and so on. Generally, the family of thin film electronic devices includes thin film transistors (TFTs), thin film solar cells (TFSCs), thin film sensors (TFSs), thin film memories (TFMs), and many other conventional and novel devices. Although these devices belong to different industrial fields, they share many common material physics, device theories, fabrication methods, and application fundamentals. Therefore, a collection of studies on these devices may broaden our knowledge about this advancing field and lead to some novel ideas about the related research works. Accordingly, the Special Issue "Recent Advances in Thin Film Electronic Devices"in *Micromachines*, the collection of this reprint, is an approach.

There are 1 editorial and 11 papers are included in this reprint. Device fundamentals, fabrication processes, and testing methods of thin film electronic devices are covered. The experimental data, simulation results, and theoretical analysis presented in these studies should push the advances in thin film electronic devices as well as the research activities employed by the researchers in this field.

Special thanks to Editor Tu and the other members of *Micromachines*! With their strong support, this reprint has become possible.

Chengyuan Dong
Editor

 micromachines

MDPI

Editorial

Editorial for the Special Issue on Recent Advances in Thin Film Electronic Devices

Chengyuan Dong

Department of Electronic Engineering, Shanghai Jiao Tong University, Shanghai 200240, China; cydong@sjtu.edu.cn

Citation: Dong, C. Editorial for the Special Issue on Recent Advances in Thin Film Electronic Devices. *Micromachines* **2022**, *13*, 1445. https://doi.org/10.3390/mi13091445

Received: 29 August 2022
Accepted: 30 August 2022
Published: 1 September 2022

Publisher's Note: MDPI stays neutral with regard to jurisdictional claims in published maps and institutional affiliations.

Thin film electronic devices have been attracting more and more attention because of their applications in many industry fields, such as in flat panel displays (FPDs), energy devices, sensors, memories, and so on [1–3]. From a fabrication point of view, thin film electronic devices can not only be prepared on rigid substrates (including glass, wafers, etc.) but also on flexible substrates (including polymers, paper, etc.); this makes give thin film electronic devices the potential to be used in some quickly advancing fields, such as the Internet of Things and medical electronics [4,5].

The family of thin film electronic devices includes thin film transistors (TFTs), thin film solar cells (TFSCs), thin film sensors (TFSs), thin film memories (TFMs), and many other conventional and novel devices. To fabricate them, many advanced preparation methods, including magnetron sputtering, chemical vapor deposition (CVD), atomic layer deposition (ALD), lithography, dry etching, as well as solution methods, are employed. To improve the yield for mass productions, some effective testing methods are widely used.

This Special Issue comprises 11 original papers about recent advances in the research and development of thin film electronic devices. Specifically, three research fields are covered: device fundamentals (five papers), fabrication processes (five papers), and testing methods (one paper). These typical studies reveal the recent advances in thin film electronic devices, which are briefly summarized as follows.

Defect density dominates the electrical properties of semiconductor films and devices. In this Special Issue, J. C. Tinoco et al. investigated the impact of the semiconductor defect density on solution-processed flexible Schottky barrier diodes (SBDs) [6]. The simulation analysis and experimental measurements confirmed that it was necessary to consider the presence of a density of states in the semiconductor gap for standard SBDs to understand specific changes observed in their performance.

Nitrogen-doping is an effective method to improve the electrical performance and stability of amorphous InGaZnO (a-IGZO) TFTs. A technology computer-aided design (TCAD) simulation was employed to analyze the nitrogen-doping effect on sub-gap density of states in a-IGZO TFTs [7]. The numerical simulation results displayed that the interface trap states, bulk tail states, and deep-level sub-gap defect states originating from oxygen-vacancy-related defects might be effectively suppressed by an appropriate nitrogen-doping treatment.

In addition to TFTs, the gate-all-around field-effect transistors (GAA FETs) were also simulated to find the mechanisms about reducing power with punch-through current annealing [8]. To maximize power efficiency during electro-thermal annealing, the application of gate module engineering was confirmed to be more suitable than the isolation or source drain modules.

It is interesting that some novel functions could be realized by $CsPbI_3$ thin films. J. Y. Chen et al. confirmed the learning and memory behavior similar to biological neurons in Au/$CsPbI_3$/ITO structure [9]; they also discussed the possibility of forming long-term memory in the device through changing input signals.

In recent years, many researchers have proposed novel photonic crystal fibers (PCF)-based polarization filters. Here, S. Selvendran et al. [10] used D-shaped PCF to form a reconfigurable surface-plasmon-based filter/sensor; they achieved a maximum confinement loss of about 713 dB/cm at the operating wavelength of 1.98 μm in X-polarization by the surface plasmon effect.

Generally, electrical properties of thin films are influenced by many processing conditions. The effect of the deposition time on the structural and 3D vertical growth and electrical conductivity properties of electrodeposited anatase-rutile nanostructured thin films was studied [11], proving that the deposition time during the electrophoretic experiment consistently evidently affected the structure, morphology, and electrical conductivity of the corresponding films.

Recently, fabrication improvement in metal oxide TFTs has become a hot topic. Two papers related to this issue are included in this Special Issue [12,13]. N. Chen et al. [12] tried to apply laser treatment in the solution-processing of active layers of metal oxide TFTs, covering laser photochemical cracking of metastable bonds, laser thermal effect, photoactivation effect, and laser sintering of nanoparticles. In addition, W. Zhang et al. [13] investigated atmosphere effect in post-annealing treatments for a-IGZO TFTs with SiO_x passivation layers, where different atmospheres (air, N_2, O_2, and vacuum) were studied at length.

Interestingly, some novel processes relating thin film electronic devices were reported in this Special Issue [14,15]. K. S. Lee employed a process simplification for n-type nanosheet FETs without a ground plane region [14]; the proposed flow could be performed in situ, without the requirement of changing chambers or a high-temperature annealing process. In addition, X. Ding et al. successfully realized the efficient multi-material structured thin film transfer to elastomers for stretchable electronic devices by combining bench-top thin film structuring with solvent-assisted lift-off methods [15].

Liquid crystal displays (LCDs) are still the mainstream of FPDs, so it is very important to use effective inspection methods to improve yield in the mass productions of TFT-LCDs. Accordingly, non-contact optical detection of foreign materials adhered to color filters (CFs) and TFTs was investigated by F. M. Tzu et al. [16]. In contrast to the height of the debris material, the image was acquired by transforming the geometric shape from a square for side-view illumination using area charge-coupled devices (CCDs) in this study. The relating experiments presented a successful design to prevent a valuable component malfunction.

I would like to thank all the authors for submitting their papers to the Special Issue "Recent Advances in Thin Film Electronic Devices", as well as all the reviewers and editors for their contributions to improving these submissions.

Funding: This work was supported by the Key Research Project of Jiangxi Province (Grant No. 20194ABC28005).

Conflicts of Interest: The author declares no conflict of interest.

References

1. Kasap, S.; Capper, P. *Handbook of Electronic and Photonic Materials;* Springer International Publishing: New York, NY, USA, 2007.
2. Brotherton, S.D. *Introduction to Thin Film Transistors;* Springer International Publishing: Cham, Switzerland, 2013.
3. Zhou, Y. *Semiconducting Metal Oxide Thin-Film Transistors;* IOP Publishing Ltd.: Cham, Switzerland, 2021.
4. Someya, T. *Stretchable Electronics;* Wiley-VCH Verlag & Co., KGaA: Weinheim, Germany, 2013.
5. Rogers, J.A.; Ghaffari, B.; Kim, D.H. *Stretchable Bioelectronics for Medical Devices and Systems;* Springer International Publishing: Cham, Switzerland, 2016.
6. Tinoco, J.C.; Hernandez, S.A.; Olvera, M.L.; Estrada, M.; Garcia, R.; Martinez-Lopez, A.G. Impact of the Semiconductor Defect Density on Solution-Processed Flexible Schottky Barrier Diodes. *Micromachines* **2022**, *13*, 800. [CrossRef] [PubMed]
7. Zhu, Z.; Cao, W.; Huang, X.; Shi, Z.; Zhou, D.; Xu, W. Analysis of Nitrogen-Doping Effect on Sub-Gap Density of States in a-IGZO TFTs by TCAD Simulation. *Micromachines* **2022**, *13*, 617. [CrossRef] [PubMed]
8. Kim, M.K.; Choi, Y.K.; Park, J.Y. Power Reduction in Punch-Through Current-Based Electro-Thermal Annealing in Gate-All-Around FETs. *Micromachines* **2022**, *13*, 124. [CrossRef] [PubMed]

9. Chen, J.Y.; Tang, X.G.; Liu, Q.X.; Jiang, Y.P.; Zhong, W.M.; Luo, F. An Artificial Synapse Based on $CsPbI^3$ Thin Film. *Micromachines* **2022**, *13*, 284. [CrossRef] [PubMed]
10. Selvendran, S.; Divya, J.; Raja, A.S.; Sivasubramanian, A.; Itapu, S. A Reconfigurable Surface-Plasmon-Based Filter/Sensor Using D-Shaped Photonic Crystal Fiber. *Micromachines* **2022**, *13*, 917. [CrossRef] [PubMed]
11. Amancio, M.A.; Romaguera-Barcelay, Y.; Matos, R.S.; Pires, M.A.; Gandarilla, A.M.D.; Nascimento, M.V.B.; Nobre, F.X.; Talu, S.; Filho, D.F.F.; Brito, W.R. Effect of the Deposition Time on the Structural, 3D Vertical Growth, and Electrical Conductivity Properties of Electrodeposited Anatase–Rutile Nanostructured Thin Films. *Micromachines* **2022**, *13*, 1361. [CrossRef] [PubMed]
12. Chen, N.; Ning, H.; Liang, Z.; Liu, X.; Wang, X.; Yao, R.; Zhong, J.; Fu, X.; Qiu, T.; Peng, J. Application of Laser Treatment in MOS-TFT Active Layer Prepared by Solution Method. *Micromachines* **2021**, *12*, 1496. [CrossRef] [PubMed]
13. Zhang, W.; Fan, Z.; Shen, A.; Dong, C. Atmosphere Effect in Post-Annealing Treatments for Amorphous InGaZnO Thin-Film Transistors with SiOx Passivation Layers. *Micromachines* **2021**, *12*, 1551. [CrossRef] [PubMed]
14. Lee, K.S.; Park, J.Y. Article N-Type Nanosheet FETs without Ground Plane Region for Process Simplification. *Micromachines* **2022**, *13*, 432. [CrossRef] [PubMed]
15. Ding, X.; Moran-Mirabal, M. Efficient Multi-Material Structured Thin Film Transfer to Elastomers for Stretchable Electronic Devices. *Micromachines* **2022**, *13*, 334. [CrossRef] [PubMed]
16. Tzu, F.M.; Hsu, S.H.; Chen, J.S. Non-Contact Optical Detection of Foreign Materials Adhered to Color Filter and Thin-Film Transistor. *Micromachines* **2022**, *13*, 101. [CrossRef]

MDPI

Article

Impact of the Semiconductor Defect Density on Solution-Processed Flexible Schottky Barrier Diodes

Julio C. Tinoco [1,2], Samuel A. Hernandez [1], María de la Luz Olvera [3], Magali Estrada [3], Rodolfo García [4] and Andrea G. Martinez-Lopez [1,2,*]

1 Micro and Nanotechnology Research Centre (MICRONA), Universidad Veracruzana, Veracruz 94294, Mexico; jutinoco@uv.mx (J.C.T.); samuhernandez@uv.mx (S.A.H.)
2 Facultad de Ingeniería de la Construcción y el Hábitat (FICH), Universidad Veracruzana, Veracruz 94294, Mexico
3 Solid-State Electronics Section, Electrical Engineering Department, CINVESTAV-IPN, Mexico City 07360, Mexico; molvera@cinvestav.mx (M.d.l.L.O.); mestrada@cinvestav.mx (M.E.)
4 University Center UAEM Ecatepec, Universidad Autónoma del Estado de México, Ecatepec de Morelos 55020, Mexico; rzgarcial@uaemex.mx
* Correspondence: andmartinez@uv.mx

Abstract: Schottky barrier diodes, developed by low-cost techniques and low temperature processes (LTP-SBD), have gained attention for different kinds of novel applications, including flexible electronic fabrication. This work analyzes the behavior of the *I–V* characteristic of solution processed, ZnO Schottky barrier diodes, fabricated at a low temperature. It is shown that the use of standard extraction methods to determine diode parameters in these devices produce significant dispersion of the ideality factor with values from 2.2 to 4.1, as well as a dependence on the diode area without physical meaning. The analysis of simulated *I–V* characteristic of LTP-SBD, and its comparison with experimental measurements, confirmed that it is necessary to consider the presence of a density of states (DOS) in the semiconductor gap, to understand specific changes observed in their performance, with respect to standard SBDs. These changes include increased values of *Rs*, as well as its dependence on bias, an important reduction of the diode current and small rectification values (*RR*). Additionally, it is shown that the standard extraction methodologies cannot be used to obtain diode parameters of LTP-SBD, as it is necessary to develop adequate parameter extraction methodologies for them.

Keywords: zinc oxide films; solution-processing electronics; Schottky barrier diodes; semiconductor defects

Citation: Tinoco, J.C.; Hernandez, S.A.; Olvera, M.d.l.L.; Estrada, M.; García, R.; Martinez-Lopez, A.G. Impact of the Semiconductor Defect Density on Solution-Processed Flexible Schottky Barrier Diodes. *Micromachines* **2022**, *13*, 800. https://doi.org/10.3390/mi13050800

Academic Editor: Chengyuan Dong

Received: 28 April 2022
Accepted: 19 May 2022
Published: 21 May 2022

1. Introduction

During the last decades, the microelectronics industry has been exploring a technological diversification, which allows the possibility of developing specific electronic systems for nontraditional areas, like medical and health care systems, environmental, biological applications, detection systems for chemical or physical signals, etc. Further development of novel materials and fabrication methodologies is required to produce new devices with the desired features. Some of these applications can require substrates, like flexible, transparent, organic, paper, among others [1]. In this context, film deposition from precursor solutions appears to be a potential tool for novel materials and electronic device fabrication techniques [2,3].

On another hand, Schottky barrier diodes (SBD), based on nanostructured oxide semiconductor films, appear to be potential candidates for different kinds of sensor devices. Furthermore, the possibility of using solution-processing techniques for diode manufacture allows the reduction of the fabrication temperatures to levels which make the full fabrication process compatible with flexible substrates. In recent years, the development of SBD based on solution-processes has become an interesting technological approach for manufacturing flexible and paper-based electronic devices, including a variety of sensor devices.

Up to now, for the analysis of low-cost and low temperature processed SBD, (LTP-SBD), thermionic emission was considered the main conduction mechanism. The diode parameters, such as barrier height (ϕ_b), ideality factor (η), and series resistance (R_S), are obtained using parameter extraction methodologies developed for high quality, crystalline semiconductor-based SBD, processed at high temperatures.

In LTP-SBD, three main features have been observed in diode parameters, extracted using above mentioned methodologies [4–11]: (i) the rectification ratio ($RR = I_{ON}/I_{OFF}$) is, usually, very small (one or two orders of magnitude); (ii) large ideality factor values ($\eta > 2$), and (iii) relatively large series resistances (R_S). Regarding η, values close to 2, even greater than 7, can be found in the literature [4–11]. For SBD processed using high vacuum techniques, like sputtering deposition at room temperature, values of η near to 1 have been found [12,13].

Trying to understand the differences observed in the extracted diode parameters for LTP-SBD, with respect to those obtained for standard SBD, different explanations have been considered, among which are the impact of R_S, the presence of different conduction mechanisms, barrier height inhomogeneities, and interfacial states [5–7]. However, the physical reasons behind the wide range of values obtained for the LTP-SBD ideality factor are not clear, nor are the strong differences in parameter values observed in different film deposition methods.

In addition to the possible causes of this observed behavior, low temperature processing could jeopardize the semiconductor film quality, since it is well known that noncrystalline materials present a density of localized states (DOS) within the energy gap, which can strongly affect the behavior of devices based on these materials.

In this work, the behavior of the *I–V* characteristic of solution processed, ZnO LTP-SBD is studied. Main diode parameters, obtained by four extraction methods used for standard SBD fabricated at higher temperatures, are analyzed to evaluate the possibility of using them to characterize LTP-SBD. Additionally, simulated *I–V* characteristic of SBD, considering the presence of a density of localized states inside the semiconductor gap, were obtained to analyze the origin of the main characteristics of diodes performance.

2. Experimental Part

2.1. Fabrication Process

ZnO Schottky barrier diodes were obtained as follows: (i) the synthesis of ZnO nanoparticles; (ii) the deposition of a film, consisting of the ZnO nanoparticle colloidal dispersion, on a polyethylene terephthalate (PET) substrate, covered by an indium tin oxide (ITO) film as back-side electrode; (iii) deposition, by screen-printing technique, of a top silver electrode. The fabrication process was limited to a maximum temperature of 150 °C. A detailed fabrication process can be found in [11]. Devices with square shape and different length (L) were manufactured and then electrically characterized.

2.2. SBD Simulation

SBDs were simulated using the ATLAS simulation program from Silvaco [14]. ZnO was considered the semiconductor material and the presence of DOS was included.

As already mentioned, noncrystalline semiconductor materials contain certain distributions of DOS, which dominate the overall device electrical characteristics. Such states are grouped into deep and tail states. For our study, the effect of the tail states is predominant, so we will only consider them in our simulations. Tail state energy distribution can be approximated to an exponential distribution as:

$$g(E) = N_{TA} exp\left(-\frac{E_C - E}{E_{TA}}\right) + N_{TD} exp\left(-\frac{E - E_V}{E_{TD}}\right) \quad (1)$$

where N_{TA} and N_{TD} are, respectively, the acceptor and donor density of the tail states at the corresponding band border. E_{TA} and E_{TD} are, respectively, the activation energy of the acceptor and donor tails.

Main material parameters are shown in Table 1. For the DOS, a symmetrical variation for the acceptor and donor tail states was considered, while the value of N_{TA} and N_{TD}, was varied from 10^{18} to 10^{20} cm^{-3} eV^{-1}. For E_{TA} and E_{TD}, typical values for metal oxide materials were considered. Two different carrier densities, N_B, were analyzed. The metal work function (Φ_M) was fixed to produce a barrier height of 0.54 eV.

Table 1. Summary of the semiconductor film parameters used in simulations.

Parameter	Value	Parameter	Value
E_g	3.2 eV	N_{TA}, N_{TD}	10^{18} to 10^{20} eV^{-1} cm^{-3}
χ_s	4.3 eV	E_{TA}	0.105 eV
N_C, N_V	5×10^{18} cm^{-3}	E_{TD}	0.385 eV
μ_e	10 cm^2/Vs	N_B	5×10^{16} and 5×10^{18} cm^{-3}

3. Traditional Extraction Methods to Obtain SBD Main Parameters from I–V Curves

Considering that the thermionic emission is the main conduction mechanism in SBD, four *I–V* extraction methodologies have been used to determine the diode parameters.

3.1. Ideal Extraction Method

The diode current (I_D) of an ideal SBD diode is defined by the general diode equation (GDE) expressed as:

$$I_D = I_0\left[exp\left(\frac{qV_D}{\eta kT}\right) - 1\right] \quad (2)$$

where V_D is the applied voltage and η is the ideality factor.

The term I_0 is the reverse current, which is defined as:

$$I_0 = AA^*T^2exp\left(-\frac{q}{kT}\phi_b\right) \quad (3)$$

where A is the device area, A^* is the Richardson constant and ϕ_b is the barrier height formed between the metal and the semiconductor.

There are different procedures to determine the barrier height and the ideality factor. Combining (2) and (3), and considering $V_D >> kT/q$, the GDE can be expressed as:

$$ln(I_D) = ln\left(AA^*T^2\right) - \frac{q}{kT}\phi_b + \frac{q}{\eta kT}V_D \quad (4)$$

Therefore, the semilogarithmic plot of the forward characteristic exhibits a linear dependence where the slope is related to η and the y-axis intercept with ϕ_b.

3.2. Norde's Function

This method considers the presence of a resistance in series with an ideal diode (R_S). The method was developed without considering the ideality factor [15]. Afterwards η was included into the extraction procedure [16].

Considering the series resistance, the GDE is modified as:

$$I_D = I_0\left[exp\left(\frac{q(V_D - I_D R_s)}{\eta kT}\right) - 1\right] \quad (5)$$

This method is based on the definition of an F function, considering $V_D >> kT/q$, as [16]:

$$F(V_D,\gamma) = \frac{V_D}{\gamma} - \frac{kT}{q}ln\left(\frac{I_D}{AA^*T^2}\right) \quad (6)$$

where γ is an arbitrary constant greater that η.

For values of γ greater than η, the $F(V_D,\gamma)$ vs. V_D plot presents a minimum at the point (V_0, F_0). That point corresponds with a diode current, of value I_0 [16]. Hence, the ideality factor can be determined, considering two different values of γ (γ_1 and γ_2) and the corresponding diode current values at the minimum of the F function [16]:

$$\eta = \frac{\gamma_1 I_{02} - \gamma_2 I_{01}}{I_{02} - I_{01}} \tag{7}$$

The barrier height and the series resistance can be determined as [16]:

$$\phi_b = F_{01} + \left(\frac{1}{\eta} - \frac{1}{\gamma_1}\right)V_{01} - \frac{kT}{q}\frac{\gamma_1 - \eta}{\eta} = F_{02} + \left(\frac{1}{\eta} - \frac{1}{\gamma_2}\right)V_{02} - \frac{kT}{q}\frac{\gamma_2 - \eta}{\eta} \tag{8}$$

$$R_s = \frac{kT}{q}\frac{\gamma_1 - \eta}{I_{01}} = \frac{kT}{q}\frac{\gamma_2 - \eta}{I_{02}} \tag{9}$$

3.3. Cheung's Function

The Cheung´s method allows the determination of the ideality factor, as well as the series resistance and the barrier height [17]. Equation (5) can be rewritten, considering the current density ($J_D = I_D/A$) and $V_D >> kT/q$, as:

$$V_D = R_s A J_D + \eta\phi_B + \frac{kT}{q}\eta \cdot ln\left(\frac{J_D}{A^*T^2}\right) \tag{10}$$

The derivative of Equation (10) with respect to the logarithm of the current density is defined as [17]:

$$\frac{dV_D}{dln(J_D)} = R_s A J_D + \frac{kT}{q}\eta \tag{11}$$

As can be seen, a linear dependence with J_D is present and the ideality factor can be determined from the corresponding *y*-axis intercept [17].

Moreover, an H function is defined as [16]:

$$H(J_D) \equiv V_D - \frac{kT}{q}\eta \cdot ln\left(\frac{J_D}{A^*T^2}\right) \tag{12}$$

Comparing Equations (10) and (12), the H function is expressed as [17]:

$$H(J_D) = R_s A J_D + \eta\phi_B \tag{13}$$

Therefore, the plot of H vs. J_D shows a linear behavior. The slope is related to the series resistance and the *y*-axis intercept to the barrier height.

3.4. Forward–Reverse (F–R) Function

This method analyses both the forward as well as reverse I–V characteristic. According to Equation (3), the reverse current is bias independent. However, solution-processed devices usually exhibit an important variation of the reverse current with the reverse voltage [4–11]. Considering the image force lowering effect and a very thin semiconductor film, the reverse current density (J_R) can be expressed as [11]:

$$J_R = A^*T^2 exp\left[-\frac{q}{kT}\left(\phi_b - \sqrt{\frac{qV_R}{4\pi\varepsilon_0 k_d t}}\right)\right] \tag{14}$$

where V_R is the reverse bias, k_d is the dynamic dielectric constant and t is the electrical semiconductor film thickness.

Hence, ϕ_b can be determined through the *y*-axis intercept of the semilogarithmic plot of J_R vs. ${V_R}^{1/2}$.

Furthermore, the series resistance can be extracted by the voltage derivative, with respect to the diode current of the forward characteristic (dV_F/dI_F), which is expressed as [18]:

$$\frac{dV_F}{dI_F} = \left(\frac{dI_F}{dV_F}\right)^{-1} = R_S + \eta\frac{kT}{q}\left(\frac{1}{I_F + I_0}\right) \quad (15)$$

Therefore, R_S can be extracted from the y-axis intercept of the plot of the inverse of the current derivative vs. the $1/(I_F + I_0)$ term [11]. I_0 is calculated through (3) using the ϕ_b value obtained from (14). Additionally, η can be extracted, in an independent manner, from the slope of the same plot [11].

4. Results and Discussion

Figure 1 shows the J_D–V_D characteristics of measured devices with different L. As can be seen, the rectification ratio RR, obtained as diode current at +1 V divided by the current at −1 V, is about one order of magnitude. Contrarily, for crystalline devices, RR can be greater than six orders of magnitude. Moreover, the reverse current density shows a significant dependence on the applied voltage. On the other hand, as expected, the forward current density has similar values for all devices at low applied voltage. Beyond 0.5 V, J_D starts to increase as the device area is reduced.

Diode parameter extractions, using the different strategies explained in Section 3, were performed with the aim of analysing the diode performance, as well as the unforeseen current density increment, to a deeper level. Table 2 summarizes the extracted values for ϕ_b, η and R_S. The overall results agree with the main features previously observed for LTP-SBD [4–11].

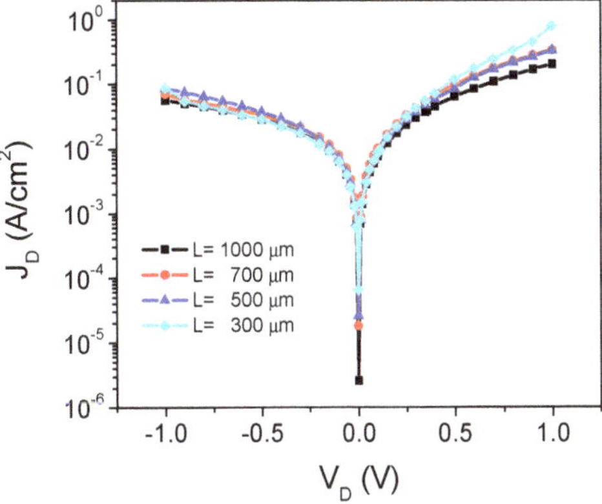

Figure 1. Plot of the J_D–V_D characteristic for the different length devices. The diode area is defined as $A = L^2$.

Table 2. Summary of the extracted Schottky barrier diode through the different extraction procedures.

Diode Length (µm)	Ideal Method		Norde's Function			Cheung's Method			F–R Method		
	η	ϕ_b (eV)	η	ϕ_b (eV)	R_s (Ω)	η	ϕ_b (eV)	R_s (Ω)	η	ϕ_b (eV)	R_s (Ω)
1000	16.5	0.48	2.2	0.53	592	2.4	0.54	506	4.1	0.52	395
700	15.6	0.47	2.6	0.55	714	3.1	0.53	546	3.2	0.52	675
500	14.4	0.48	3.0	0.56	1478	3.1	0.53	1189	3.1	0.52	1300
300	12	0.48	3.5	0.59	2300	2.3	0.54	3238	3	0.53	3300

Figure 2 shows the comparison of the extracted barrier height obtained with the different methods. As can be seen, for all devices the extracted values of ϕ_b using the ideal method are about 10% smaller than using the other methods. The Cheung and the F–R methods exhibit close values of ϕ_b, while Norde´s method shows a small difference for large devices, which increases as L is reduced. Nevertheless, the different methods used allow the determination of the barrier height with a relatively small variation of $\pm$10 %. It is worth noting that the Norde and Cheung methods allow the extraction of the barrier height after the ideality factor; hence, a reliable η extraction is of main importance. On the contrary, the F–R method allows the determination of the barrier height in an independent form.

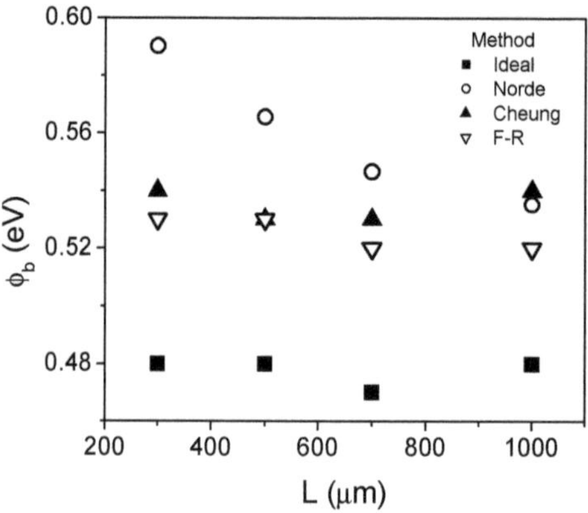

Figure 2. Comparison of the extracted values of the barrier height using the different methods.

As can be seen from Table 2, the ideality factor presents abnormally large values when the ideal extraction is used. This occurs because the impact on R_S is neglected. The other methods include the series resistance and, thus, the extracted values of η are reduced. Figure 3 shows the comparison of the η extracted values using the methods which include the series resistance on the extraction methodology. In all cases, the resulting η values are greater than 2 and a significant dispersion is observed. Additionally, Norde and F–R methods show an opposite trend. In the Norde case, the value of η increases as the device area is reduced, while with the F–R method, it reduces. However, there is not a physical reason that could support the variation of the diode ideality factor with the device area.

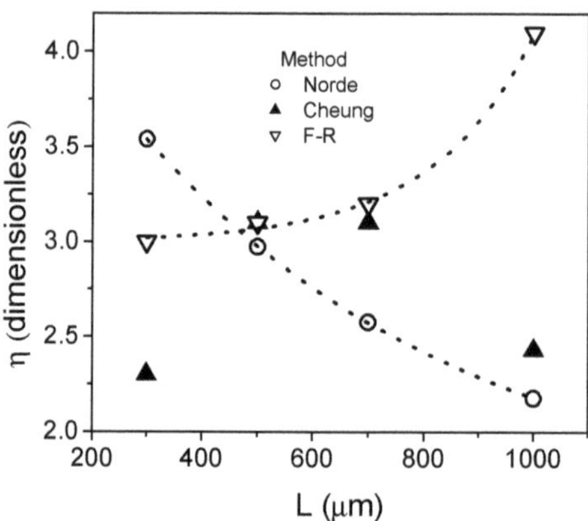

Figure 3. Comparison of the extracted values of the ideality factor.

Therefore, the standard SBD extraction methods allow the determination of the barrier height into a reasonable deviation. For the ideality factor, however, a significant dispersion and even different trends when varying the diode area are obtained, depending on the method considered. However, the behavior shown in Figure 1 suggests that a single set of parameters is required to define the diode performance up to ~0.4 V, and after that bias, the increment on the current density with the area must be explained.

Figure 4 shows the extracted R_S vs. the inverse of the diode area. In the inset, the R_S vs. L plot is shown. In general, R_S has a linear behavior in respect to $1/A$. As a first approach, the device can be considered a rectangular semiconductor with electrodes on the top and at the bottom, which produce the observed dependence with the diode area. This fact also suggests a constant value of the resistance normalized with diode area ($A.R_S$).

The above-mentioned analysis implies an issue on the proper determination of the diode parameters utilizing the traditional methodologies, since a single set of parameters cannot explain the experimental diode current. Because of the observed results, a deeper analysis on the LTP-SBD behavior must be performed, as well as the development of specific extraction methodologies for this kind of devices.

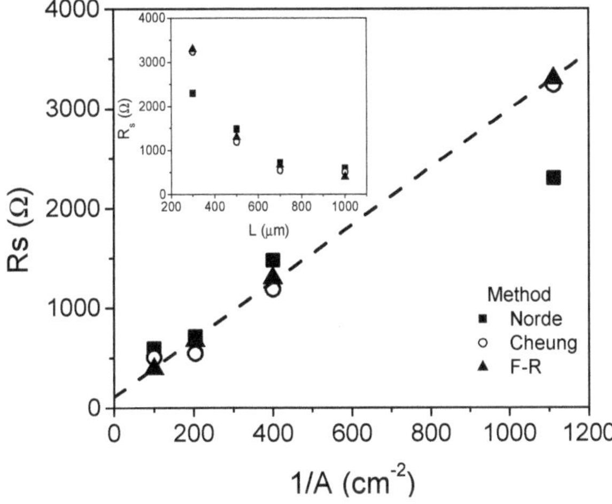

Figure 4. Comparison of the extracted values of the series resistance using the different methods.

To further analyze the behavior of LTP-SBDs, finite-element numerical simulations were performed. Figure 5a,b show the *I–V* characteristic for devices simulated in ATLAS, for two doping concentrations and the different DOS parameters shown in Table 1. For comparison, defect-free simulated devices are included. As can be seen, the reverse current exhibits a negligible impact with the presence of DOS. Contrarily, the tail state's presence produces an important reduction in the forward diode current. This can be explained due to the electron-trapping on the defects, which implies a reduction in the overall free carriers in the conduction band and, hence, of the device current. The impact, however, is more important for devices with relatively low free carrier concentration, which implies a highly resistive film. In such cases, I_{ON} is reduced by several orders of magnitude, which explains the typical rectification ratio experimentally achieved [4–11].

Figure 6 shows the comparison of the *RR* vs. the defect densities for two values of N_B. As can be observed, LTP-SBDs with low quality semiconductor films (i.e., high resistivity and high defects densities) exhibit a strong *RR* reduction, until around one order of magnitude, which are similar to what is observed in experimental devices [4–11]. On the other hand, when the layer has a moderate or high conductivity (which implies a better film quality), the impact of the defects is reduced and the *RR* value is only slightly reduced, remaining several orders of magnitude as experimentally observed for high vacuum processing [12,13], even for defect densities in the range of 10^{20} cm^{-3}eV^{-1}. It can be expected that a further increase of the defect densities will produce a stronger *RR* reduction.

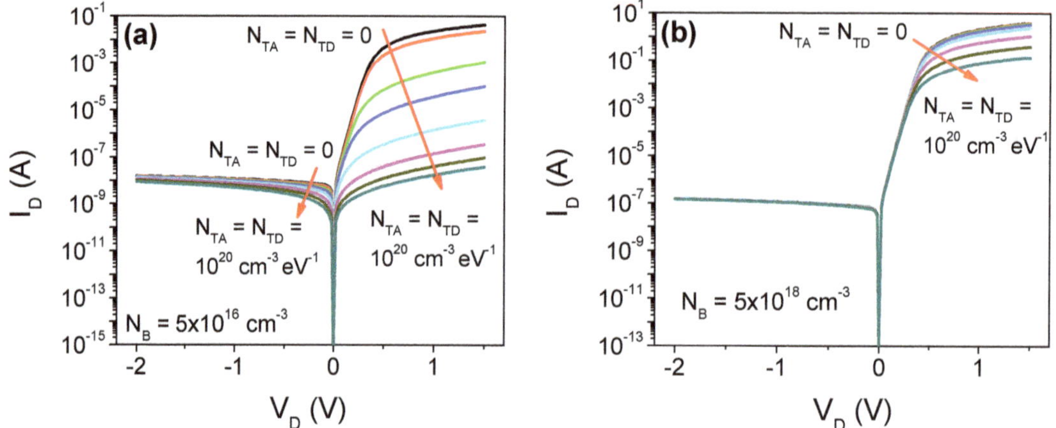

Figure 5. *I–V* characteristic for the simulated SBD considering different densities of localized states for N_B equal to (**a**) 5×10^{16} and (**b**) 5×10^{18} cm^{-3}. For comparison, defect free devices ($N_{TA} = N_{TD} = 0$) are considered.

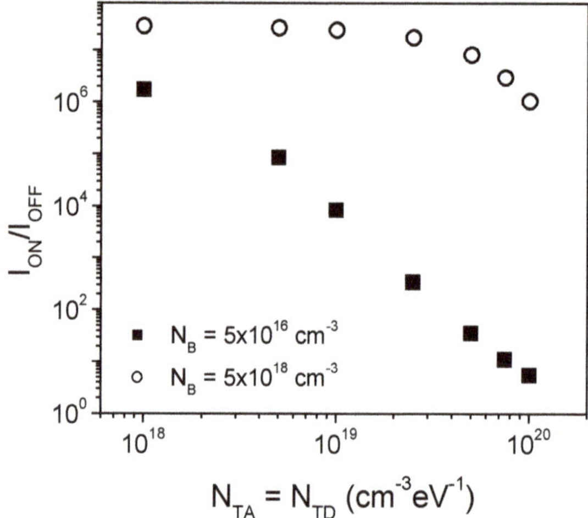

Figure 6. Comparison of the rectification ratio (*RR*) vs. the defect densities, for both N_B values used in the simulations.

Figure 7 shows the simulated electron concentration (n_e) inside the semiconductor film for both N_B values vs. forward applied voltage. As can be seen, the impact of the film qualities on n_e is confirmed. Moreover, it is observed that, for high resistivity films, n_e is modulated by the forward bias. Contrarily, for low resistivity films, n_e is constant along most of the film thickness.

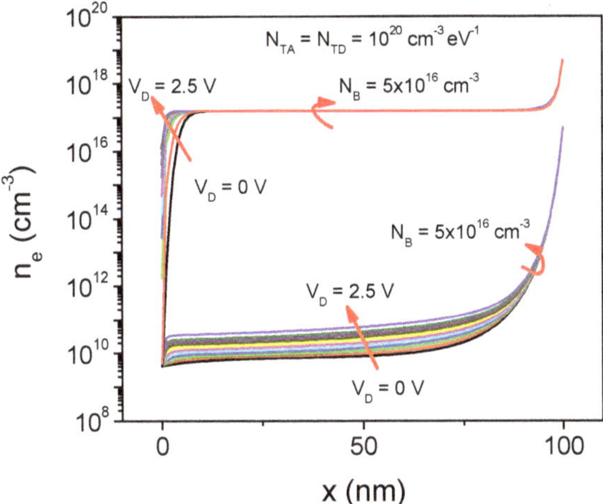

Figure 7. Comparison of the electron concentration (n_e) vs. the semiconductor film position (x), for both N_B values used in the simulations.

As was mentioned above, the diode can be considered as a rectangular semiconductor die. Therefore, the resistance due to a differential film thickness is defined as:

$$dR_S = \frac{1}{q\mu A} \cdot \frac{dx}{n_e(x)} \tag{16}$$

where μ is the electron mobility and A is the device area.

The total resistance, due to the semiconductor film, can be calculated integrating (16) along the film thickness (t):

$$R_S = \frac{1}{q\mu A} \int_0^t \frac{dx}{n_e(x)} \tag{17}$$

Considering the electron distribution shown in Figure 7, it is possible to determine the series resistance contribution caused by the film. Figure 8 shows the calculated resistance vs. the bias applied for device with $N_B = 5 \times 10^{16}$ cm^{-3} and defect densities of 10^{20} cm^{-3} eV^{-1}. In the inset, the calculated resistance for a device with $N_B = 5 \times 10^{18}$ cm^{-3} is also shown. For the case of low resistivity semiconductor film, the resistance shows an abrupt reduction at small forward bias, from about 1 MΩ to few kΩ. When bias is increased above 1 V, R_S becomes almost constant. On the contrary, for a higher resistivity material with a relatively high defect density, the resistance exhibits extremely high values. At the same time, an important dependence on the applied forward voltage is observed. According to Figure 8, R_S exhibits an exponential dependence on V_D, which, as a first approach, can be expressed as:

$$R_S = R_0 exp\left(-\frac{V_D}{\delta}\right) \tag{18}$$

where R_0 is the zero bias resistance and δ can be related to the DOS.

Under this scenario, to better represent the behavior of the I–V curve for LTP-SBDs, the general diode equation must be modified as:

$$I_D = AA^*T^2 exp\left(-\frac{q\phi_B}{kT}\right) exp\left\{\frac{q\left[V_D - I_D R_0 exp\left(-\frac{V_D}{\delta}\right)\right]}{\eta kT}\right\} \tag{19}$$

This series resistance bias dependence can explain the abnormal current density increment observed as L is reduced in Figure 1. As diode area is reduced, the resistance value is increased, and, therefore, its reduction with the applied voltage becomes more significant. Hence, beyond 0.4 V, the current density starts to increase, due to the R_S reduction.

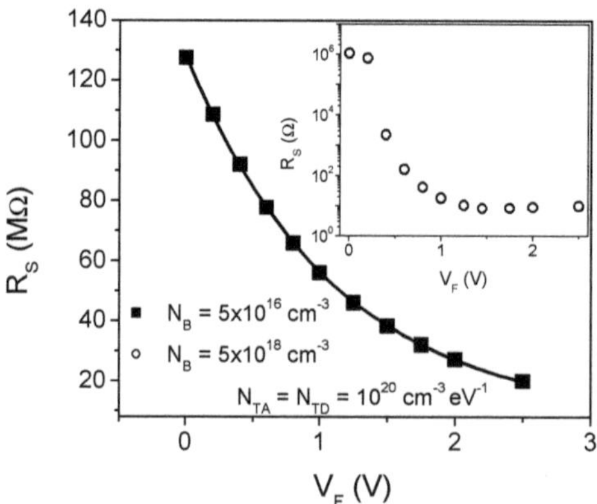

Figure 8. Calculated series resistance vs. forward bias, for N_B of 5×10^{16} cm^{-3}. In the inset, the corresponding plot for N_B of 5×10^{18} cm^{-3} is shown.

Furthermore, the bias-dependent R_S would imply an important concern regarding the correct parameter extraction. In order to verify this assumption, Figure 9 shows the corresponding $dV_D/dln(I_D)$ vs. I_D plot, according to Equation (11) of Cheung´s method, for the simulated device shown in Figure 8. For comparison, in the inset the plot for a defect free device is shown. As can be seen, the extraction procedure can be properly applied for the defect free device getting the extracted value of η as one. On the contrary, when the high defect density is included in the simulation, the plot does not show a linear behavior at any forward bias region. This fact clearly shows that for low-cost and low-temperature processing SBDs, the film quality compromises the reliable application of the traditional extraction methods due to the bias dependence exhibited by R_S. Therefore, the extracted parameters can exhibit the important variations shown in Figures 2 and 3. Thus, proper extraction methodologies for low-cost and low-temperature processed SBD are of main importance to adequately understand the diode behavior.

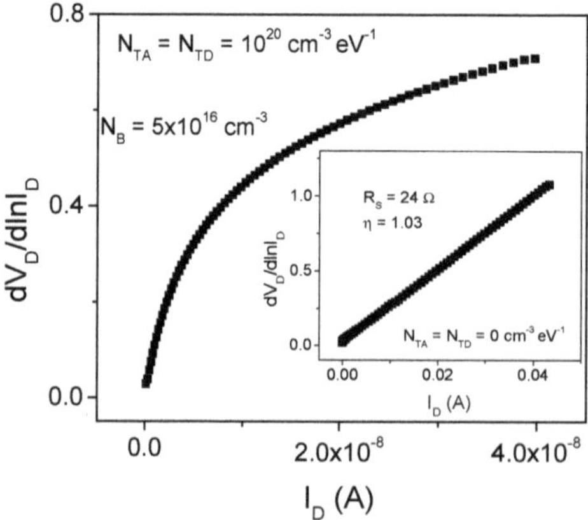

Figure 9. Plot of the $dV_D/dln(I_D)$ vs I_D, used for η and R_S extraction in the Cheung extraction method.

5. Conclusions

ZnO *LTP-SBDs* were analyzed using a single I_D–V_D characteristic and four traditional extraction methodologies. The barrier height extraction shows a relatively small dispersion of about ±10%. On the other hand, the ideality factor obtained exhibits a significant dispersion with values from 2.2 to 4.1, depending on the extraction method used. Simulation results show that devices without or with low DOS, as is the case of standard SBD fabricated at higher temperatures, show high values of RR, relatively small values of R_S, and are almost bias-independent at relatively high forward applied voltage. Thus, traditional parameter extraction methodologies can be properly used. On the other hand, devices fabricated using low-cost techniques (solution-processing techniques, printing strategies, etc.) at low temperatures can produce films with high tail state densities and high resistivity. Simulations showed that the combination of high DOS and low carrier concentration produces a strong impact on diode behavior, which implies an important reduction in the forward current and *RR* values. For these devices, the series resistance exhibits high values, as well as an exponential dependence on the forward applied voltage. Under these conditions, the traditional extraction methodologies of diode parameters are compromised, so further efforts must be made to develop adequate parameter extraction methodologies for low-cost and very low-temperature processed Schottky barrier diodes.

Author Contributions: Conceptualization, J.C.T. and A.G.M.-L.; methodology, J.C.T. and A.G.M.-L.; software, J.C.T., R.G. and M.E.; validation, S.A.H. and A.G.M.-L.; investigation, J.C.T., S.A.H., R.G. and A.G.M.-L.; resources, M.d.l.L.O., M.E. and A.G.M.-L.; writing—original draft preparation, J.C.T. and A.G.M.-L.; writing—review and editing, M.E. All authors have read and agreed to the published version of the manuscript.

Funding: This research received no external funding.

Conflicts of Interest: The authors declare no conflict of interest.

References

1. Baeg, K.-J.; Lee, J. Flexible Electronic Systems on Plastic Substrates and Textiles for Smart Wearable Technologies. *Adv. Mater. Technol.* **2020**, *5*, 2000071. [CrossRef]
2. Berger, P.R.; Li, M.; Mattei, R.M.; Niang, M.A.; Talisa, N.; Tripepi, M.; Harris, B.; Bhalerao, S.R.; Chowdhury, E.A.; Winter, C.H.; et al. Advancements in Solution Processable Devices using Metal Oxides For Printed Internet-of-Things Objects. In Proceedings of the IEEE Electron Devices Technology and Manufacturing Conference (EDTM), Singapore, 12–15 March 2019. [CrossRef]
3. Chan, K.-Y.; Ng, Z.-N.; Au, B.W.-C.; Knipp, D. Visibly transparent metal oxide diodes prepared by solution processing. *Opt. Mater.* **2018**, *75*, 595–600. [CrossRef]
4. Rouis, A.; Hizem, N.; Kalboussi, A. Electrical characterizations of Schottky diode with zinc oxide nanowires. In Proceedings of the IEEE International Conference on Design & Test of Integrated Micro & Nano-Systems (DTS), Gammarth, Tunisia, 28 April–1 May 2019. [CrossRef]
5. Ahmeda, M.A.M.; Mwankemwa, B.S.; Carleschi, E.; Doyle, B.P.; Meyer, W.E.; Nel, J.M. Effect of Sm doping ZnO nanorods on structural optical and electrical properties of Schottky diodes prepared by chemical bath deposition. *Mater. Sci. Semicond. Processing* **2018**, *79*, 53–60. [CrossRef]
6. Caglar, Y.; Caglar, M.T.; Ilican, S. XRD, SEM, XPS studies of Sb doped ZnO films and electrical properties of its based Schottky diodes. *Optik* **2018**, *164*, 424–432. [CrossRef]
7. Rana, V.S.; Rajput, J.K.; Pathak, T.K.; Purohit, L.P. Cu sputtered Cu/ZnO Schottky diodes on fluorine doped tin oxide substrate for optoelectronic applications. *Thin Solid Film.* **2019**, *679*, 79–85. [CrossRef]
8. Zhang, J.; Zhao, G.; Li, Y.; Ai, T.; Wu, C.; Jia, J.; Yan, F.; Wang, Y. Study on the electrical properties of nano ZnO/PET-ITO heterojunction prepared by hydrothermal method. *J. Electron Spectrosc. Relat. Phenom.* **2019**, *235*, 68–72. [CrossRef]
9. Ozório, M.S.; Nascimento, M.R.; Vieira, D.H.; Nogueira, G.L.; Martin, C.S.; Lima, S.A.M.; Alves, N. AZO transparent electrodes grown in situ during the deposition of zinc acetate dihydrate onto aluminum thin film by spray pyrolysis. *J. Mater. Sci. Mater. Electron.* **2019**, *30*, 13454–13461. [CrossRef]
10. Yilmaz, M.; Cirak, B.B.; Aydogan, S.; Grilli, M.L.; Biber, M. Facile electrochemical-assisted synthesis of TiO_2 nanotubes and their role in Schottky barrier diode applications. *Superlattices Microstruct.* **2018**, *113*, 310–318. [CrossRef]
11. Tinoco, J.C.; Hernández, S.A.; Rodrigez-Bernal, O.; Vega-Poot, A.G.; Rodriguez Gattorno, G.; de la L. Olvera, M.; Martinez-Lopez, A.G. Fabrication of Schottky Barrier Diodes based on ZnO for Flexible Electronics. *J. Mater. Sci. Mater. Electron.* **2020**, *30*, 7373–7377. [CrossRef]

12. Zhang, J.; Wang, H.; Wilson, J.; Ma, X.; Jin, J.; Song, A. Room Temperature Processed Ultrahigh-Frequency Indium–Gallium–Zinc-Oxide Schottky Diode. *IEEE Electron Device Lett.* **2016**, *37*, 389–392. [CrossRef]
13. Xin, Q.; Yan, L.; Du, L.; Zhang, J.; Luo, Y.; Wang, Q.; Song, A. Influence of sputtering conditions on room-temperature fabricated InGaZnO-based Schottky diodes. *Thin Solid Film.* **2016**, *616*, 569–572. [CrossRef]
14. *ATLAS User's Manual*; SILVACO Inc.: Santa Clara, CA, USA, 2020.
15. Norde, H. A modified forward I–V plot for Schottky diodes with high series resistance. *J. Appl. Phys.* **1979**, *50*, 5052–5053. [CrossRef]
16. Bohlin, K.E. Generalized Norde plot including determination of the ideality factor. *J. Appl. Phys.* **1986**, *60*, 1223–1224. [CrossRef]
17. Cheung, S.K.; Cheung, N.W. Extraction of Schottky diode parameters from forward current-voltage characteristics. *Appl. Phys. Lett* **1989**, *49*, 85–87. [CrossRef]
18. Ortiz-Conde, A.; García-Sánchez, F.J. A new approach to the extraction of single exponential diode model parameters. *Solid State Electron.* **2018**, *144*, 33–38. [CrossRef]

micromachines

MDPI

Article

Analysis of Nitrogen-Doping Effect on Sub-Gap Density of States in a-IGZO TFTs by TCAD Simulation

Zheng Zhu [1], Wei Cao [1], Xiaoming Huang [1,*], Zheng Shi [2], Dong Zhou [3] and Weizong Xu [3]

1 College of Integrated Circuit Science and Engineering, Nanjing University of Posts and Telecommunications, Nanjing 210023, China; 1219023314@njupt.edu.cn (Z.Z.); 1220024107@njupt.edu.cn (W.C.)
2 School of Communications and Information Engineering, Nanjing University of Posts and Telecommunications, Nanjing 210023, China; shizheng@njupt.edu.cn
3 National Laboratory of Solid State Microstructures, Nanjing University, Nanjing 210093, China; dongzhou@nju.edu.cn (D.Z.); njuphyxwz@126.com (W.X.)
* Correspondence: huangxm@njupt.edu.cn

Abstract: In this work, the impact of nitrogen doping (N-doping) on the distribution of sub-gap states in amorphous InGaZnO (a-IGZO) thin-film transistors (TFTs) is qualitatively analyzed by technology computer-aided design (TCAD) simulation. According to the experimental characteristics, the numerical simulation results reveal that the interface trap states, bulk tail states, and deep-level sub-gap defect states originating from oxygen-vacancy- (V_o) related defects can be suppressed by an appropriate amount of N dopant. Correspondingly, the electrical properties and reliability of the a-IGZO TFTs are dramatically enhanced. In contrast, it is observed that the interfacial and deep-level sub-gap defects are increased when the a-IGZO TFT is doped with excess nitrogen, which results in the degeneration of the device's performance and reliability. Moreover, it is found that tail-distributed acceptor-like N-related defects have been induced by excess N-doping, which is supported by the additional subthreshold slope degradation in the a-IGZO TFT.

Keywords: a-IGZO TFTs; sub-gap states; nitrogen-doping; numerical simulation; stability

Citation: Zhu, Z.; Cao, W.; Huang, X.; Shi, Z.; Zhou, D.; Xu, W. Analysis of Nitrogen-Doping Effect on Sub-Gap Density of States in a-IGZO TFTs by TCAD Simulation. *Micromachines* **2022**, *13*, 617. https://doi.org/10.3390/mi13040617

Academic Editor: Chengyuan Dong

Received: 6 March 2022
Accepted: 12 April 2022
Published: 14 April 2022

1. Introduction

Currently, the backplane technology of amorphous InGaZnO (a-IGZO) thin-film transistors (TFTs) is attracting great attention for its use in pixel switching and driving units for next-generation display applications. The competitive advantage of a-IGZO TFT technology is that it can offer high current-driving capacity, high optical transparency, low power consumption and low process temperature compared with traditional Si-based TFTs [1–3]. Although a-IGZO TFT technology has made remarkable progress since it was first proposed by Nomura et al. in 2004, these devices still cannot achieve the desired performance and reliability due to high-density sub-gap states existing in the bandgap of a-IGZO [4,5]. It has been demonstrated that the sub-gap defects mainly originate from oxygen-vacancy-related (V_o-related) defects induced by the structural disorder in a-IGZO [5–7], which degrades the electrical properties and reliability of TFTs by trapping electrons or holes in the channel layer and interfacial region under bias, light, and thermal stress [8–11]. To enhance the device performance and reliability, an in-situ nitrogen-doping (N-doping) approach during the a-IGZO active layer deposition has been proposed to suppress V_o defect generation [12–14]. For example, it has been demonstrated that N-doping can significantly improve the reliability of a-IGZO TFTs under positive gate-bias stress (PBS) and PBS with light illumination, since the N incorporated into a-IGZO will occupy the V_o sites and suppress V_o-related defect generation [14,15]. Moreover, it has also been reported that the device performance and reliability are simultaneously enhanced by N and H co-doping, which is ascribed to the passivation of the V_o distributed at the active layer and interface region by forming N—H and Zn–N bonds [16]. Although V_o-related defects can be efficiently passivated by N-doping, a fundamental physical understanding of the impact of N-doping on the distribution of sub-gap states in a-IGZO TFTs is lacking. Since the device performance and reliability basically depend upon the nature and density of sub-gap defect states [4,17], an in-depth systematic study of the impact of N-doping on the sub-gap density of states (DOS) in a-IGZO TFTs is the key to future process improvement and optimization.

In this work, the influence of N-doping on the sub-gap V_o-related defects in a-IGZO TFTs is qualitatively analyzed using technology computer-aided design (TCAD) simulation [18]. It is found that the density of the interface V_o trap states, bulk V_o-related tail states and deep-level V_o-related defect states of a-IGZO TFTs are significantly decreased by moderate N-doping, which is validated by the improvement in electrical properties and stability during PBS and sub-band illumination. In contrast, the DOS of a-IGZO TFT is increased when the a-IGZO TFTs are doped with excessive nitrogen atoms, which causes degeneration of the device performance and reliability. Meanwhile, it is confirmed that the tail-distributed acceptor-like N-related defects are formed by excessive N-doping, which leads to the degradation of subthreshold slope (*SS*) in a-IGZO TFT.

2. Experiments and Modeling Scheme

The a-IGZO TFT structure used for numerical simulation is shown in Figure 1a. The devices in this work were fabricated on n-type Si substrate. First, the gate insulator was composed of a 200 nm SiO_2 thin film grown by plasma-enhanced chemical vapor deposition (PECVD) with a rate of ~50 nm/min at 350 °C. The 45-nm-thick a-IGZO thin films were then deposited by direct-current (DC) sputtering system with a various gas mixture of $N_2/(O_2 + N_2)$ = 0%, 20%, and 40% at a fixed Ar flow rate of 30 sccm. The composition of the ceramic target used was In:Ga:Zn = 2:2:1 in atomic ratio. Subsequently, the device active region was patterned by conventional photolithography and wet chemical etching. Next, the Ti/Au (30/70 nm) bi-layer drain/source contact electrodes were evaporated by e-beam evaporation, and is the active region was further patterned using lift-off technique, which resulted in the final device dimensions of W/L = 100 μm/20 μm. Finally, a 100-nm-thick SiO_2 passivation layer was deposited by PECVD. The fabricated devices were annealed in ambient air for 1 h at 300 °C.

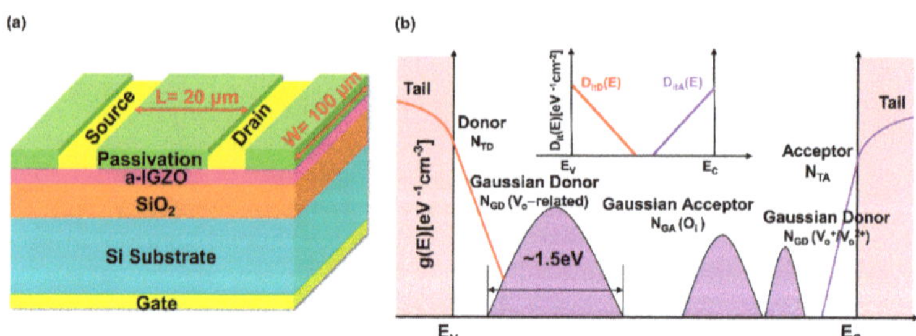

Figure 1. (**a**) Schematic diagram of the a-IGZO TFT with a bottom-gate structure; (**b**) Schematic illustration of the DOS model in the a-IGZO TFTs. The N_{TD} and N_{TA} represent the donor-like and acceptor-like tail states, respectively. The three gaussian curves represent the deep V_o-related states ($N_{GD}(V_o\text{-}related)$), oxygen interstitials ($N_{GA}(O_i)$), and shallow donor states ($N_{GD}(V_o^{+}/V_o^{2+})$), respectively. The inset is the schematic illustration of the interface trap density ($D_{it}(E)$).

In the Silvaco TCAD Simulation tool, ATLAS, a physics-based device simulator, is used to perform the electrical characterization, which can reduce the cost and time needed for experimentation [19]. It is also a powerful tool to predict the electrical behavior of specified semiconductor structures by using the Poisson and the continuity equations, which describe the electronic phenomena and electrical transport mechanism [19]. Based on the DOSs model of the a-IGZO TFTs, the types of the sub-gap states in the TFT channel region and interface region are illustrated in Figure 1b. In the a-IGZO material, the sub-gap states are mainly classified as acceptor-like and donor-like states, which can be depicted by Gaussian distribution states and exponentially decaying band-tail states. The specific mathematical model is expressed as follows [20–22]:

$$g_{TA}(E) = N_{TA} \exp\left(\frac{E - E_C}{W_{TA}}\right) \tag{1}$$

$$g_{TD}(E) = N_{TD} \exp\left(\frac{E_V - E}{W_{TD}}\right) \tag{2}$$

$$g_{GA}(E) = N_{GA} \exp\left[-\left(\frac{E_{GA} - E}{W_{GA}}\right)^2\right] \tag{3}$$

$$g_{GD}(E) = N_{GD} \exp\left[-\left(\frac{E - E_{GD}}{W_{GD}}\right)^2\right] \tag{4}$$

where the $g_{TD}(E)$ and $g_{TA}(E)$ denote the density of donor-like tail and acceptor-like tail states. The $g_{GA}(E)$ and $g_{GD}(E)$ represent the Gaussian-distributed acceptor-like and donor-like states. The N_{TA} and N_{TD} are the effective density at the conduction band minimum (E_C) and valence band maximum (E_V), respectively. The W_{TD} and W_{TA} are the characteristic slope energy of valence-band tail states and conduction-band tail states. The N_{GD} and N_{GA} are the total density of Gaussian donor and acceptor states, respectively. The E_{GA} and E_{GD} are the corresponding peak energy. W_{GA} and W_{GD} are the corresponding characteristic decay energy.

In addition, the interface trap density ($D_{it}(E)$) at the a-IGZO/dielectric interfacial region can be described as [23]:

$$D_{it}(E) = D_{itA} \exp\left(\frac{E - E_C}{W_{itA}}\right) + D_{itD} \exp\left(\frac{E_V - E}{W_{itD}}\right) \tag{5}$$

where D_{itD} and D_{itA} represent the donor-like and acceptor-like interface trap density, respectively. The W_{itA} and W_{itD} denote the corresponding slope energy.

3. Results and Discussion

Figure 2 shows the simulated and experimental transfer characteristics of the a-IGZO TFTs under various N-doping conditions at V_{DS} = 5 V. The simulation results, in consistency with the experimental data, were achieved by calibrating the D_{itA}, N_{TA}, $N_{GA}(O_i)$ and $N_{GD}(V_o{}^+/V_o{}^{2+})$, and the simulation parameters are extracted and summarized in Table 1. The total trap density of the 20% N-doping ratio a-IGZO TFT was significantly decreased compared to the undoped a-IGZO TFT. For example, the D_{itA} is decreased from 2.5×10^{13} eV^{-1} cm^{-2} to 8.0×10^{12} eV^{-1} cm^{-2}, and the N_{TA} was reduced from 8.0×10^{19} eV^{-1} cm^{-3} to 1.0×10^{19} eV^{-1} cm^{-3}. Correspondingly, the subthreshold slope (SS) was decreased from 0.8 V/dec to 0.6 V/dec, and the threshold voltage (V_{th}) was reduced from 5.0 V to 3.8 V. It has been demonstrated that the interface states and bulk traps in a-IGZO TFTs mainly originate from V_o-related defects [5,15,24]. Therefore, the simulation results confirm that the improved electrical properties of a-IGZO TFTs can be ascribed to the suppression of the generation of V_o-related defects in the device channel and interface region by N-doping. In contrast, when the N-doping ratio was increased to 40%, the number of total trap states was increased compared to the 20% N-doping ratio TFT, as shown in Table 1. Meanwhile, it was observed that the SS and V_{th} of the 40% N-doping ratio device were increased to 0.9 V/dec and 7 V, respectively, which indicates that V_o-related defects generate when the device is subjected to excessive N-doping. This result can be explained by the fact that the formation of N–Ga bonds is facilitated by heavy N-doping, which then suppresses the bonding of Ga–O in the a-IGZO thin films [14,25].

In addition, based on the simulation results, it was found that although the sub-gap DOS (N_{TA}, N_{TD}, N_{GD}(V_o-*related*), $N_{GA}(O_i)$, and $N_{GD}(V_o{}^+/V_o{}^{2+})$) existing in the 40% N-doping ratio TFT was significantly higher than that of the 20% N-doping ratio TFT, the sub-gap DOS other than N_{TA} was lower than that of the undoped TFT. Because the sub-band-gap density in a-IGZO film mainly originates from V_o-related defects, the amount of V_o in annealed a-IGZO thin films with various N-doping ratios was analyzed by X-ray photoelectron spectroscopy (XPS). Figure 3a–c show the O 1 s XPS spectra of the a-IGZO films grown using different N-doped ratios. The binding energies were calibrated by taking the C 1s as reference at 284.6 eV. Gaussian fitting was applied to decompose the combined O 1s peak. The sub-peaks centered at binding energies of 529.7 eV, 530.5 eV, and 531.5 eV were attributed to O^{2-} ions surrounded by metal atoms (In, Ga and Zn), oxygen vacancies (V_o), and OH^- impurities, respectively [26–28]. The relative level of V_o in a-IGZO film can be estimated by the proportion of peak area V_o to the whole O 1s (O_{whole}). It was found that the area proportion of V_o/O_{whole} was decreased from 35% for N-free a-IGZO film to 25% for a-IGZO film with the 20% N-doped ratio, as shown in Figure 3a,b, indicating that V_o decreases when the N is incorporated into the a-IGZO film. However, as shown in Figure 3c, it was found that the V_o increased to 31% for the a-IGZO film deposited with the 40% N-doped ratio, which means that the additional V_o was created when excess N atoms were doped into the a-IGZO film. Furthermore, the N 1s spectra XPS of the annealed a-IGZO film with 20% N-doped ratio was also analyzed, as shown in Figure 3d. The N 1 s spectrum is decomposed into two peaks at 395.7 and 397.3 eV, which are associated with the Ga Auger and N-Ga bonding [29], respectively. Therefore, the XPS results reveal that moderate N doping in a-IGZO film can suppress the generation of V_o, and excess N incorporation into a-IGZO film leads to an increase in V_o. Because the V_o existing in the 40% N-doping ratio a-IGZO film was lower than that of undoped a-IGZO film, the increased N_{TA} in 40% N-doping ratio a-IGZO TFT should be the result of the generation of N-related defects by excess N-doping [16,30], which agrees well with the increased *SS* from 0.8 V/dec to 0.9 V/dec compared to undoped a-IGZO TFT. Meanwhile, to quantitatively estimate the N concentration in the a-IGZO active layer, the actual level of N-doping in annealed a-IGZO films is characterized by secondary ion mass spectrometry (SIMS) measurement [31–33]. Figure 4 shows the depth profile of N concentration in the a-IGZO film deposited with the 20% and 40% N-doped ratios. N is clearly detectable, and there is a considerable amount of incorporated nitrogen (~10^{20} cm^{-3}) in the a-IGZO film. It has been reported that the value of the sub-gap density of states (DOSs) near the VBMs is about 5–9 × 10^{20} cm^{-3} in a-IGZO film [5,34]. In this work, it is found that when the concentration of N doping in the channel region of a-IGZO TFT with a 20% N doping ratio is ~1.0 × 10^{20} cm^{-3}, the electrical performance and stability of the device are dramatically improved. But when the concentration of N-doping in the channel region of a-IGZO TFT with a 40% N doping ratio is increased to ~1.2 × 10^{20} cm^{-3}, the electrical performance and stability of a-IGZO TFT are degraded.

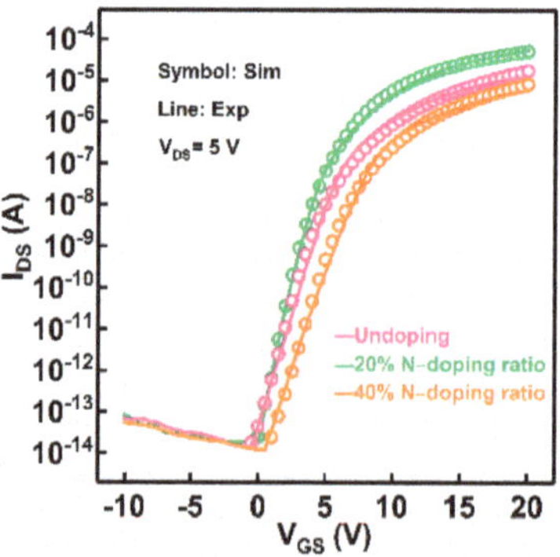

Figure 2. Simulated transfer characteristics for a-IGZO TFTs with different N-doping conditions: undoped, 20% N-doping ratio, and 40% N-doping ratio.

Table 1. Densities of key defect model parameters for a-IGZO TFT fitted after different N-doping ratios.

Parameters	Undoping	20% N-Doping Ratio	40% N-Doping Ratio	Description
D_{itA} ($eV^{-1}\ cm^{-2}$)	2.5×10^{13}	8.0×10^{12}	1.5×10^{13}	Acceptor-like interface trap densities
D_{itD} ($eV^{-1}\ cm^{-2}$)	3.0×10^{13}	9.0×10^{12}	2.0×10^{13}	Donor-like interface trap densities
N_{TA} ($eV^{-1}\ cm^{-3}$)	8.0×10^{19}	1.0×10^{19}	1.5×10^{20}	Acceptor-like tail states at E = Ec
N_{TD} ($eV^{-1}\ cm^{-3}$)	1.5×10^{20}	8.0×10^{19}	1.3×10^{20}	Donor-like tail states at E = Ev
N_{GD}(V_o-related) ($eV^{-1}\ cm^{-3}$)	8.0×10^{20}	5.0×10^{20}	6.5×10^{20}	Peak of V_o-related states
N_{GA}(O_i) ($eV^{-1}\ cm^{-3}$)	2.6×10^{17}	1.4×10^{17}	2.1×10^{17}	Peak of O_i states
N_{GD}(V_o^+/V_o^{2+}) ($eV^{-1}\ cm^{-3}$)	8.0×10^{16}	5.0×10^{16}	6.5×10^{16}	Peak of V_o^+/V_o^{2+} states

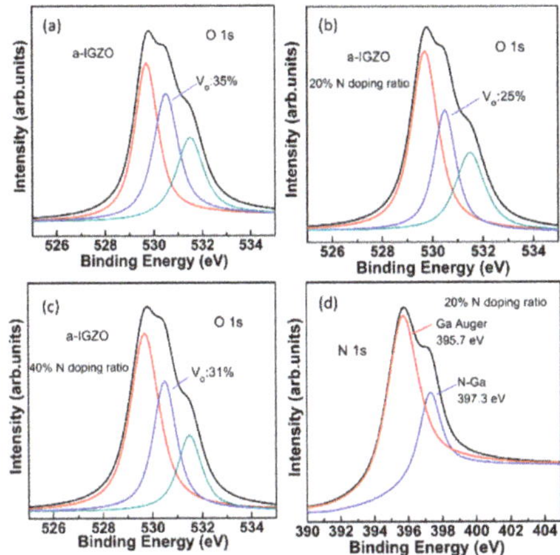

Figure 3. O 1s XPS spectra of the annealed a-IGZO films grown using N-doping ratio of (**a**) undoped, (**b**) 20% and (**c**) 40%. (**d**) N 1s XPS spectra of a-IGZO film grown with 20% N-doping ratio.

According to the distribution of the sub-gap DOS fitted in a-IGZO TFTs with various N-doping ratios, a comprehensive quantitative study on the device stability under positive bias stress (PBS) was carried out. During the PBS process, the TFTs were applied at a V_{GS} of 15 V for the stress duration of 5000 s. Figure 5a–c show the experimental and simulated evolution of transfer characteristics as a function of PBS time for the a-IGZO TFTs with different N-doping ratios. It was found that the shift in threshold voltage (ΔV_{th}) induced by PBS was 2.06 V, 0.8 V, and 1.68 V for undoped a-IGZO TFT, 20% N-doping a-IGZO TFT, and 40% N-doping a-IGZO TFT, respectively. It has been reported that the shift in V_{th} (ΔV_{th}) of a-IGZO TFTs under PBS basically originates from the interfacial V_o-related defects trapping electrons at the device interfacial region [35,36]. In the simulation results, it was clearly seen that the $D_{it}(E)$ for the 20% N-doping ratio a-IGZO TFT was lower than of the undoped a-IGZO TFT and 40% N-doping ratio a-IGZO TFT. For example, the D_{itA} for undoped a-IGZO TFT, 20% N-doped a-IGZO TFT, and 40% N-doped a-IGZO TFT is 2.5×10^{13} eV^{-1} cm^{-2}, 8.0×10^{12} eV^{-1} cm^{-2}, and 1.5×10^{13} eV^{-1} cm^{-2}, respectively, suggesting that interfacial V_o-related defects can be suppressed by moderate N-doping.

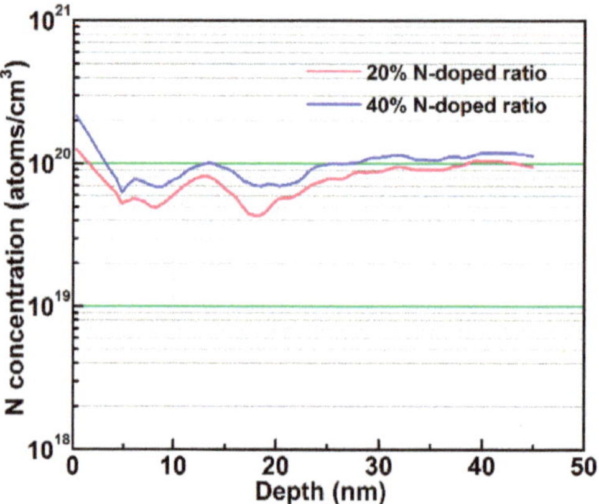

Figure 4. Depth profile of nitrogen in a-IGZO film deposited under 20% N-doping ratio and 40% N-doping ratio.

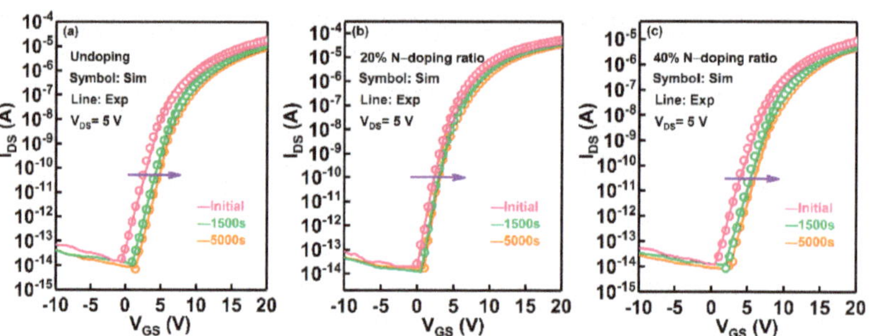

Figure 5. Simulated transfer characteristics against positive bias stress (PBS) time for the a-IGZO TFTs fabricated with different N-doping ratios: (**a**) undoped, (**b**) 20% N-doping ratio, and (**c**) 40% N-doping ratio.

In addition, it has been reported that weak oxygen ions originating from structural disorder in a-IGZO TFTs cause ΔV_{th} under PBS [37]. During the PBS process, the weak oxygen ions are ionized to generate oxygen interstitials (O_i) because of their low formation energies [4,38,39]. Meanwhile, according to the first-principle studies, the generated O_i during PBS forms an octahedral configuration [O_i(oct)] and is electrically active. Correspondingly, the introduced O_i(oct)-related defect states are distributed above the mid-gap (E_i) in the a-IGZO TFTs [40]. When the Fermi level moves up under PBS, the O_i(oct)-related states are filled by trapping electrons and thus negatively charged to generate O_i^{2-}oct) [41]. Figure 6a–c show the forming process of O_i^{2-}(oct)-related charged states in a-IGZO TFTs undergoing PBS. Because of the structural relaxation effect, the O_i^{2-}(oct)-related charged states are transformed into deep-level negative-U states, which are located below the mid-gap in the a-IGZO TFTs [37,42]. As a result, although new O_i^{2-}(oct)-related charged states are generated under the PBS process, the *SS* of a-IGZO TFTs has no apparent change due to the negative-U property of the O_i^{2-}(oct) states.

According to the simulation results, $N_{GA}(O_i)$ exhibited an increasing trend for a-IGZO TFTs with various N-doping ratios under the PBS process, as shown in Table 2. For example, the $N_{GA}(O_i)$ in the N-free a-IGZO TFT continuously increased from 2.6×10^{17} eV^{-1} cm^{-3} to 3.6×10^{17} eV^{-1} cm^{-3} after 5000 s PBS. This result shows that the O_i(oct)-related defects were generated in the a-IGZO TFT upon PBS, originating from weak oxygen ions in the device channel region. Compared to the undoped a-IGZO TFT, the generated O_i(oct)-related trap states under PBS in the 20% N-doping ratio a-IGZO TFT decreased from 3.6×10^{17} eV^{-1} cm^{-3} to 2.5×10^{17} eV^{-1} cm^{-3} after 5000 s PBS, due to the suppression of V_o-related traps by N-doping. Correspondingly, the device had superior electrical reliability under PBS. In contrast, the generated O_i(oct)-related trap states under PBS in the 40% N-doping ratio a-IGZO TFT (3.2×10^{17} eV^{-1} cm^{-3}) were higher than that of the 20% N-doping ratio a-IGZO TFT, due to the V_o-related traps generated by heavy N-doping, resulting in device-stability degeneration under PBS.

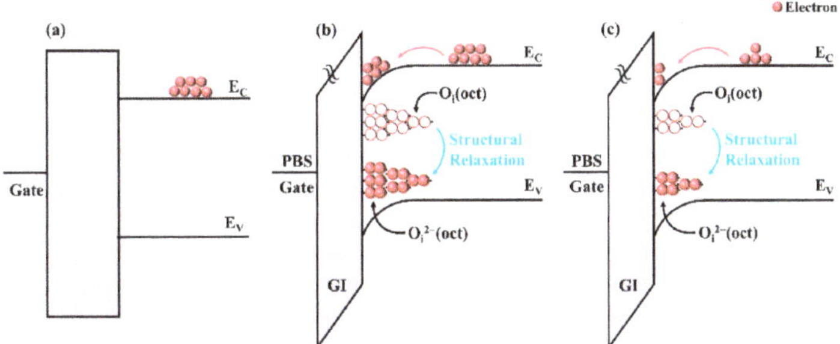

Figure 6. Schematic of the energy-band diagram of the a-IGZO TFTs. (**a**) the energy-band diagram of the TFTs before PBS, (**b**,**c**) the energy-band diagrams of the undoped TFTs and 20% N-doping ratio TFTs during PBS, respectively.

Table 2. Densities of key defect model parameters for a-IGZO TFT fitted with different N-doping ratios after PBS.

Parameters	N-Doping Ratio	Initial	1500 s	5000 s	Description
$N_{GA}(O_i)$ (eV^{-1} cm^{-3})	0%	2.6×10^{17}	3.2×10^{17}	3.6×10^{17}	Peak of O_i states
	20%	1.4×10^{17}	2.0×10^{17}	2.5×10^{17}	
	40%	2.1×10^{17}	2.8×10^{17}	3.2×10^{17}	

Finally, to reveal the impact of N-doping on the distributions of deep-level sub-gap states, the transfer characteristics of a-IGZO TFTs with various N-doping conditions under sub-band-gap light illumination were simulated. The simulation results are shown in Figure 7a–c, and the simulation parameters are extracted in Table 3. It was found that the I-V curves of the undoped a-IGZO TFT exhibited an overall shift in a negative direction with the decrease in incident light wavelength from 650 nm to 500 nm, as shown in Figure 7a. The negative ΔV_{th} is attributable to the photorelease of occupied electrons in the interface states and deep-level sub-gap states. It has been demonstrated that the deep-level defects in a-IGZO mainly originate from neutral V_o, which would be entirely occupied above E_V with an energy width of ~1.5 eV [5,24,43]. As a result, the occupied interfacial and deep-level V_o-related defects would be ionized into V_o^+/V_o^{2+} under the corresponding photon energy illumination, which agrees well with the simulation result that the $N_{GD}(V_o^+/V_o^{2+})$ exhibits continuously increase as the illumination wavelength decreases [44,45], as shown in Table 3. It is clear that the $N_{GD}(V_o^+/V_o^{2+})$ of the undoping a-IGZO TFT is increased from 1.5×10^{17} eV^{-1} cm^{-3} to 3.0×10^{17} eV^{-1} cm^{-3} with the decrease of incident light wavelength from 650 nm to 500 nm. In addition, based on the first-principle calculation and experimental observation, the activation energy (E_a) of the photoexcited ionization process from occupied deep level V_o to V_o^+ and V_o^{2+} is required to be ~2.0 eV and ~2.3 eV, respectively [37,45]. Meanwhile, these photo-induced transitions (both V_o to V_o^+ and V_o to V_o^{2+}) could cause the outward relaxation in the vicinity of metal atoms, which lead to the generation of new defect level near the E_i and E_c edge [24,45]. The formation process of V_o^+ and V_o^{2+} states induced by sub-band-gap illumination is illustrated in Figure 8. In the simulation, it is observed that the generated $N_{GA}(V_o^+\text{-}related)$ near the mid-gap is 9.0×10^{16} eV^{-1} cm^{-3} at λ = 600 nm (~2.0 eV), and the generated $N_{GD}(V_o^{2+}\text{-}related)$ near bottom of the conduction band is 1.2×10^{17} eV^{-1} cm^{-3} at λ = 500 nm (~2.3 eV). It has been reported that the variation of SS (ΔSS) is in connection with the amount of created trap states (ΔN_t) in the TFTs channel and interface region, which is expressed by using the following equation [45]:

$$\Delta SS = \frac{\Delta N_t \ln(10)kT}{C_i} \tag{6}$$

where k is the Boltzmann's constant, T is the absolute temperature, C_i is the capacitance of the gate dielectric. Therefore, the SS degradation for the undoping a-IGZO TFT induced by incident illumination of $\lambda \leq 600$ nm is caused by the new defects creation near the E_i and E_c edge.

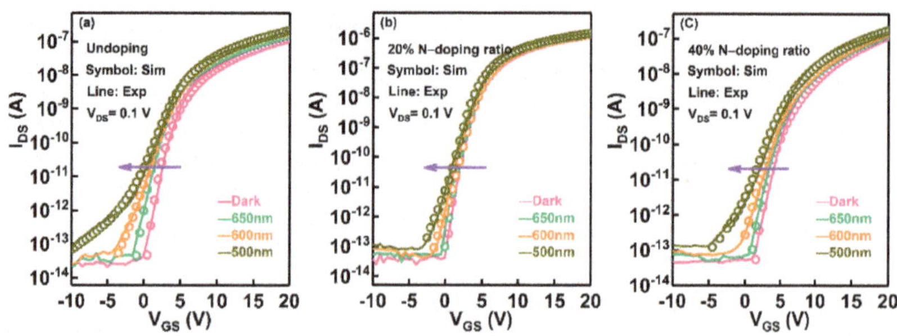

Figure 7. Simulated transfer characteristics against various monochromatic light illumination for the a-IGZO TFTs fabricated with different N-doping ratios: (**a**) undoping, (**b**) 20% N-doping ratio, and (**c**) 40% N-doping ratio.

Table 3. Densities of key defect model parameters for a-IGZO TFT fitted with different N-doping ratios after monochromatic light illumination.

Parameters	N-Doping Ratio	Dark	650 nm	600 nm	500 nm	Description
N_{GD}(V_o-related) (eV^{-1} cm^{-3})	0%	8.0×10^{20}				Peak of V_o-related states
	20%	5.0×10^{20}				
	40%	6.5×10^{20}				
N_{GD}(V_o^+/V_o^{2+}) (eV^{-1} cm^{-3})	0%	8.0×10^{16}	1.5×10^{17}	2.5×10^{17}	3.0×10^{17}	Peak of V_o^+/V_o^{2+} states
	20%	5.0×10^{16}	9.0×10^{16}	1.5×10^{17}	2.2×10^{17}	
	40%	6.5×10^{16}	1.2×10^{17}	2.0×10^{17}	2.5×10^{17}	
N_{GD}(V_o^{2+}-related) (eV^{-1} cm^{-3})	0%	—	—	—	1.2×10^{17}	Peak of V_o^{2+}-related states
	20%	—	—	—	7.0×10^{16}	
	40%	—	—	—	9.0×10^{16}	
N_{GA}(V_o^+-related) (eV^{-1} cm^{-3})	0%	—	—	9.0×10^{16}	9.0×10^{16}	Peak of V_o^+-related states
	20%	—	—	4.0×10^{16}	4.0×10^{16}	
	40%	—	—	7.5×10^{16}	7.5×10^{16}	

Comparatively, it is found that the density of deep-level V_o-related traps is suppressed by moderate N-doping into the a-IGZO TFT. As shown in Table 3, the $N_{GD}(V_o\text{-}related)$ in the 20% N-doping ratio a-IGZO TFT is decreased from 8.0×10^{20} eV^{-1} cm^{-3} to 5.0×10^{20} eV^{-1} cm^{-3} compared with the undoping a-IGZO TFT, and the generated $N_{GD}(V_o^{2+}\text{-}related)$ is reduced from 1.2×10^{17} eV^{-1} cm^{-3} to 7.0×10^{16} eV^{-1} cm^{-3} under λ= 500 nm. Correspondingly, the electrical stability of the 20% N-doping ratio a-IGZO TFT under light illumination is significantly improved. It is found that the ΔSS and ΔV_{th} in the 20% N-doping ratio a-IGZO TFTs (0.58 V/dec; −0.8 V) are lower than that of undoping a-IGZO TFT (1.95 V/dec; −1.6 V) at λ = 500 nm, as shown in Figure 7b. Therefore, it can be concluded that the improved device reliability under sub-band light illumination is owing to the passivation of V_o-related traps in the device channel region by moderate N-doping. However, the deep-level V_o-related defect in the 40% N-doping ratio a-IGZO TFT is increased to 6.5×10^{20} eV^{-1} cm^{-3} compared with the 20% N-doping ratio a-IGZO TFT, and the generated $N_{GD}(V_o{}^{2+}\text{-}related)$ is increased to 9.0×10^{16} eV^{-1} cm^{-3} under λ = 500 nm. Meanwhile, the degradation of SS and V_{th} (1.5 V/dec; −1.3 V) are observed in the 40% N-doping ratio a-IGZO TFT. This result means that superfluous N-doping into the a-IGZO TFTs will result in the increase of deep-level V_o traps [25], which degrades the device stability under sub-band-gap illumination.

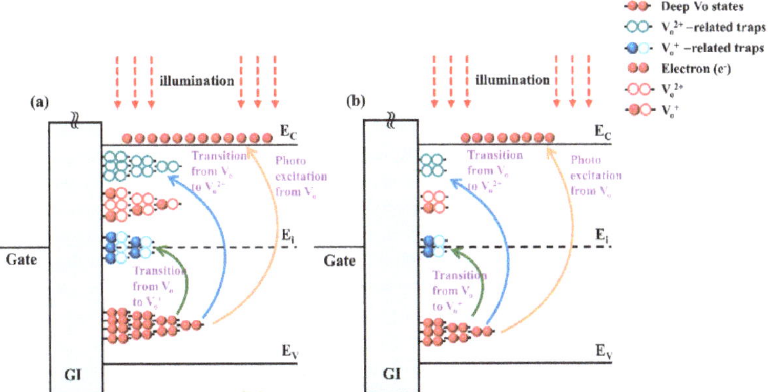

Figure 8. Schematic of the generation process of V_o-related defect states under short-wavelength light illumination: (**a**) undoping, (**b**) 20% N-doping ratio.

4. Conclusions

In this work, the fundamental physical understanding of the N-doping on DOS over the whole sub-band-gap range has been analyzed by Silvaco TCAD simulation. It is found that the improved electrical performances for the 20% N-doping ratio a-IGZO TFT are owing to the suppression of interface V_o trap states and bulk tail states (V_o-related) by N-doping. Meanwhile, the O_i and deep-level V_o-related traps are suppressed by an appropriate amount of N dopant, which causes the improvement of device stability during PBS and sub-band illumination processes by suppressing the formation of O_i and the photoexcited ionization from occupied deep-level V_o to $V_o{}^{+}$ and $V_o{}^{2+}$, respectively. In contrast, the excessive N-doping will cause the generation of acceptor-like N-related defects and the increase of V_o-related traps in the channel and interface region of a-IGZO TFTs, which leads to the degeneration of the device performance and reliability.

Author Contributions: Conceptualization, Z.Z., W.C. and X.H.; Writing—original draft, Z.Z.; Writing—review & editing, W.C., X.H., Z.S., D.Z. and W.X. All authors have read and agreed to the published version of the manuscript.

Funding: This work was supported by the National Natural Science Foundation of China (Grant No. 61904086) and China Postdoctoral Science Foundation (Grant No. SBH19006).

Conflicts of Interest: The authors declare no conflict interest.

References

1. Nomura, K.; Ohta, H.; Takagi, A.; Kamiya, T.; Hirano, M.; Hosono, H. Room-temperature fabrication of transparent flexible thin-film transistors using amorphous oxide semiconductors. *Nature* **2004**, *432*, 488–492. [CrossRef] [PubMed]
2. Kamiya, T.; Hosono, H. Material characteristics and applications of transparent amorphous oxide semiconductors. *NPG Asia Mater.* **2010**, *2*, 15–22. [CrossRef]
3. Yu, X.; Marks, T.J.; Facchetti, A. Metal oxides for optoelectronic applications. *Nat. Mater.* **2016**, *15*, 383–396. [CrossRef] [PubMed]
4. Janotti, A.; Van de Walle, C.G. Native point defects in ZnO. *Phys. Rev. B* **2007**, *76*, 165202. [CrossRef]

5. Nomura, K.; Kamiya, T.; Yanagi, H.; Ikenaga, E.; Yang, K.; Kobayashi, K.; Hirano, M.; Hosono, H. Subgap states in transparent amorphous oxide semiconductor, In–Ga–Zn–O, observed by bulk sensitive X-ray photoelectron spectroscopy. *Appl. Phys. Lett.* **2008**, *92*, 202117. [CrossRef]
6. Yao, J.; Xu, N.; Deng, S.; Chen, J.; She, J.; Shieh, H.D.; Liu, P.T.; Huang, Y.P. Electrical and Photosensitive Characteristics of a-IGZO TFTs Related to Oxygen Vacancy. *IEEE Trans. Electron Devices* **2011**, *58*, 1121–1126.
7. Fishchuk, I.I.; Kadashchuk, A.; Bhoolokam, A.; de Jamblinne de Meux, A.; Pourtois, G.; Gavrilyuk, M.M.; Köhler, A.; Bässler, H.; Heremans, P.; Genoe, J. Interplay between hopping and band transport in high-mobility disordered semiconductors at large carrier concentrations: The case of the amorphous oxide InGaZnO. *Phys. Rev. B* **2016**, *93*, 195204. [CrossRef]
8. Xiao, X.; Zhang, L.; Shao, Y.; Zhou, X.; He, H.; Zhang, S. Room-Temperature-Processed Flexible Amorphous InGaZnO Thin Film Transistor. *ACS Appl. Mater. Interfaces* **2018**, *10*, 25850–25857. [CrossRef]
9. Li, S.; Wang, M.; Zhang, D.; Wang, H.; Shan, Q. A Unified Degradation Model of a-InGaZnO TFTs Under Negative Gate Bias with or without an Illumination. *IEEE J. Electron Devices Soc.* **2019**, *7*, 1063–1071. [CrossRef]
10. Huang, X.; Wu, C.; Lu, H.; Ren, F.; Xu, Q.; Ou, H.; Zhang, R.; Zheng, Y. Electrical instability of amorphous indium-gallium-zinc oxide thin film transistors under monochromatic light illumination. *Appl. Phys. Lett.* **2012**, *100*, 243505. [CrossRef]
11. Dai, C.; Qi, G.; Qiao, H.; Wang, W.; Xiao, H.; Hu, Y.; Guo, L.; Dai, M.; Wang, P.; Webster, T.J. Modeling and Mechanism of Enhanced Performance of In-Ga-Zn-O Thin-Film Transistors with Nanometer Thicknesses under Temperature Stress. *J. Phys. Chem. C* **2020**, *124*, 22793–22798. [CrossRef]
12. Eun Kim, C.; Yun, I. Effects of nitrogen doping on device characteristics of InSnO thin film transistor. *Appl. Phys. Lett.* **2012**, *100*, 013501. [CrossRef]
13. Cheng, Y.C.; Chang, S.P.; Chen, I.C.; Tsai, Y.L.; Cheng, T.H.; Chang, S.J. Polycrystalline In–Ga–O Thin-Film Transistors Coupled with a Nitrogen Doping Technique for High-Performance UV Detectors. *IEEE Trans. Electron Devices* **2020**, *67*, 140–145. [CrossRef]
14. Park, K.; Kim, J.; Sung, T.; Park, H.; Baeck, J.; Bae, J.; Park, K.; Yoon, S.; Kang, I.; Chung, K.; et al. Highly Reliable Amorphous In-Ga-Zn-O Thin-Film Transistors Through the Addition of Nitrogen Doping. *IEEE Trans. Electron Devices* **2019**, *66*, 457–463. [CrossRef]
15. Liu, J.; Guo, J.; Yang, W.; Wang, C.; Yuan, B.; Liu, J.; Wu, Z.; Zhang, Q.; Liu, D.; Chen, H.; et al. Graded Channel Junctionless InGaZnO Thin-Film Transistors with Both High Transporting Properties and Good Bias Stress Stability. *ACS Appl. Mater. Interfaces* **2020**, *12*, 43950–43957. [CrossRef]
16. Abliz, A.; Gao, Q.; Wan, D.; Liu, X.; Xu, L.; Liu, C.; Jiang, C.; Li, X.; Chen, H.; Guo, T.; et al. Effects of Nitrogen and Hydrogen Codoping on the Electrical Performance and Reliability of InGaZnO Thin-Film Transistors. *ACS Appl. Mater. Interfaces* **2017**, *9*, 10798–10804. [CrossRef]
17. Kamiya, T.; Nomura, K.; Hosono, H. Present status of amorphous In-Ga-Zn-O thin-film transistors. *Sci. Technol. Adv. Mater.* **2010**, *11*, 044305. [CrossRef]
18. SILVACO. *Atlas User's Manual: Device Simulation Software*; SILVACO: Santa Clara, CA, USA, 2004.
19. Adaika, M.; Meftah, A.; Sengouga, N.; Henini, M. Numerical simulation of bias and photo stress on indium–gallium–zinc-oxide thin film transistors. *Vacuum* **2015**, *120*, 59–67. [CrossRef]
20. Li, G.; Abliz, A.; Xu, L.; André, N.; Liu, X.; Zeng, Y.; Flandre, D.; Liao, L. Understanding hydrogen and nitrogen doping on active defects in amorphous In-Ga-Zn-O thin film transistors. *Appl. Phys. Lett.* **2018**, *112*, 253504. [CrossRef]
21. Kim, Y.; Kim, S.; Kim, W.; Bae, M.; Jeong, H.K.; Kong, D.; Choi, S.; Kim, D.M.; Kim, D.H. Amorphous InGaZnO Thin-Film Transistors—Part II: Modeling and Simulation of Negative Bias Illumination Stress-Induced Instability. *IEEE Trans. Electron Devices* **2012**, *59*, 2699–2706. [CrossRef]
22. Billah, M.; Chowdhury, M.; Mativenga, M.; Um, J.; Mruthyunjaya, R.; Heiler, G.; Tredwell, T.; Jang, J. Analysis of Improved Performance Under Negative Bias Illumination Stress of Dual Gate Driving a- IGZO TFT by TCAD Simulation. *IEEE Electron Device Lett.* **2016**, *37*, 735–738. [CrossRef]
23. Kim, Y.; Bae, M.; Kim, W.; Kong, D.; Jung, H.K.; Kim, H.; Choi, S.; Kim, D.M.; Kim, D.H. Amorphous InGaZnO Thin-Film Transistors—Part I: Complete Extraction of Density of States Over the Full Subband-Gap Energy Range. *IEEE Trans. Electron Devices* **2012**, *59*, 2689–2698. [CrossRef]
24. Ryu, B.; Noh, H.; Choi, E.; Chang, K.J. O-vacancy as the origin of negative bias illumination stress instability in amorphous In–Ga–Zn–O thin film transistors. *Appl. Phys. Lett.* **2010**, *97*, 022108. [CrossRef]
25. Liu, P.; Chang, C.; Fuh, C.; Liao, Y.; Sze, S.M. Effects of Nitrogen on Amorphous Nitrogenated InGaZnO (a-IGZO:N) Thin Film Transistors. *J. Disp. Technol.* **2016**, *12*, 1070–1077. [CrossRef]
26. Lee, K.W.; Kim, K.M.; Heo, K.Y.; Park, S.K.; Lee, S.K.; Kim, H.J. Effects of UV light and carbon nanotube dopant on solution-based indium gallium zinc oxide thin-film transistors. *Curr. Appl. Phys.* **2011**, *11*, 280–285. [CrossRef]
27. Yang, S.; Hwan Ji, K.; Ki Kim, U.; Seong Hwang, C.; Ko Park, S.; Hwang, C.; Jang, J.; Kyeong Jeong, J. Suppression in the negative bias illumination instability of Zn-Sn-O transistor using oxygen plasma treatment. *Appl. Phys. Lett.* **2011**, *99*, 102103. [CrossRef]
28. Hernandez Gutierrez, C.A.; Casallas Moreno, Y.L.; Rangel Kuoppa, V.T.; Cardona, D.; Hu, Y.; Kudriatsev, Y.; Zambrano Serrano, M.A.; Gallardo Hernandez, S.; Lopez Lopez, M. Study of the heavily p-type doping of cubic GaN with Mg. *Sci. Rep.* **2020**, *10*, 16858. [CrossRef]
29. Jiang, Y.; Wang, Q.; Zhang, F.; Li, L.; Zhou, D.; Liu, Y.; Wang, D.; Ao, J. Reduction of leakage current by O2 plasma treatment for device isolation of AlGaN/GaN heterojunction field-effect transistors. *Appl. Surf. Sci.* **2015**, *351*, 1155–1160. [CrossRef]

30. Chien, J.F.; Chen, C.H.; Shyue, J.J.; Chen, M.J. Local electronic structures and electrical characteristics of well-controlled nitrogen-doped ZnO thin films prepared by remote plasma in situ atomic layer doping. *ACS Appl. Mater. Interfaces* **2012**, *4*, 3471–3475. [CrossRef]
31. Hernández Gutiérrez, C.A.; Kudriavtsev, Y.; Cardona, D.; Guillén Cervantes, A.; Santana Rodríguez, G.; Escobosa, A.; Hernández Hernández, L.A.; López López, M. In_xGa_{1-x} N nucleation by In+ ion implantation into GaN. *Nucl. Instrum. Methods Phys. Res. Sect. B Beam Interact. Mater. At.* **2017**, *413*, 62–67. [CrossRef]
32. Hernández Gutiérrez, C.A.; Kudriavtsev, Y.; Cardona, D.; Hernández, A.G.; Camas-Anzueto, J.L. Optical, electrical, and chemical characterization of nanostructured InxGa1-xN formed by high fluence In+ ion implantation into GaN. *Opt. Mater.* **2021**, *111*, 110541. [CrossRef]
33. Hu, Y.; Hernandez Gutierrez, C.A.; Solis Cisneros, H.I.; Santana, G.; Kudriatsev, Y.; Camas Anzueto, J.L.; Lopez Lopez, M. Blue luminescence origin and Mg acceptor saturation in highly doped zinc-blende GaN with Mg. *J. Alloys Compd.* **2022**, *897*, 163133. [CrossRef]
34. Kamiya, T.; Nomura, K.; Hosono, H. Origins of High Mobility and Low Operation Voltage of Amorphous Oxide TFTs: Electronic Structure, Electron Transport, Defects and Doping. *J. Disp. Technol.* **2009**, *5*, 468–483. [CrossRef]
35. Vygranenko, Y.; Wang, K.; Nathan, A. Stable indium oxide thin-film transistors with fast threshold voltage recovery. *Appl. Phys. Lett.* **2007**, *91*, 263508. [CrossRef]
36. Kwon, D.W.; Kim, J.H.; Chang, J.S.; Kim, S.W.; Kim, W.; Park, J.C.; Song, I.; Kim, C.J.; Jung, U.I.; Park, B. Temperature effect on negative bias-induced instability of HfInZnO amorphous oxide thin film transistor. *Appl. Phys. Lett.* **2011**, *98*, 063502. [CrossRef]
37. Ide, K.; Kikuchi, Y.; Nomura, K.; Kimura, M.; Kamiya, T.; Hosono, H. Effects of excess oxygen on operation characteristics of amorphous In-Ga-Zn-O thin-film transistors. *Appl. Phys. Lett.* **2011**, *99*, 093507. [CrossRef]
38. Zhou, X.; Shao, Y.; Zhang, L.; Lu, H.; He, H.; Han, D.; Wang, Y.; Zhang, S. Oxygen Interstitial Creation in a-IGZO Thin-Film Transistors Under Positive Gate-Bias Stress. *IEEE Electron Device Lett.* **2017**, *38*, 1252–1255. [CrossRef]
39. Chowdhury, M.D.H.; Migliorato, P.; Jang, J. Time-temperature dependence of positive gate bias stress and recovery in amorphous indium-gallium-zinc-oxide thin-film-transistors. *Appl. Phys. Lett.* **2011**, *98*, 153511. [CrossRef]
40. Omura, H.; Kumomi, H.; Nomura, K.; Kamiya, T.; Hirano, M.; Hosono, H. First-principles study of native point defects in crystalline indium gallium zinc oxide. *J. Appl. Phys.* **2009**, *105*, 093712. [CrossRef]
41. Zhou, X.; Shao, Y.; Zhang, L.; Xiao, X.; Han, D.; Wang, Y.; Zhang, S. Oxygen Adsorption Effect of Amorphous InGaZnO Thin-Film Transistors. *IEEE Electron Device Lett.* **2017**, *38*, 465–468. [CrossRef]
42. Anderson, P.W. Model for the Electronic Structure of Amorphous Semiconductors. *Phys. Rev. Lett.* **1975**, *34*, 953–955. [CrossRef]
43. Jang, J.T.; Park, J.; Ahn, B.D.; Kim, D.M.; Choi, S.J.; Kim, H.S.; Kim, D.H. Study on the photoresponse of amorphous In-Ga-Zn-O and zinc oxynitride semiconductor devices by the extraction of sub-gap-state distribution and device simulation. *ACS Appl. Mater. Interfaces* **2015**, *7*, 15570–15577. [CrossRef] [PubMed]
44. Kim, S.; Kim, S.; Kim, C.; Park, J.; Song, I.; Jeon, S.; Ahn, S.; Park, J.; Jeong, J.K. The influence of visible light on the gate bias instability of In–Ga–Zn–O thin film transistors. *Solid-State Electron.* **2011**, *62*, 77–81. [CrossRef]
45. Yang, Z.; Meng, T.; Zhang, Q.; Shieh, H.D. Stability of Amorphous Indium–Tungsten Oxide Thin-Film Transistors Under Various Wavelength Light Illumination. *IEEE Electron Device Lett.* **2016**, *37*, 437–440. [CrossRef]

Article

Power Reduction in Punch-Through Current-Based Electro-Thermal Annealing in Gate-All-Around FETs

Min-Kyeong Kim [1], Yang-Kyu Choi [2,*] and Jun-Young Park [1,*]

[1] School of Electronics Engineering, Chungbuk National University, Chungdae-ro 1, Chungbuk, Cheongju 28644, Korea; kming@chungbuk.ac.kr

[2] School of Electrical Engineering, Korea Advanced Institute of Science and Technology (KAIST), Daejeon 34141, Korea

* Correspondence: ykchoi@ee.kaist.ac.kr (Y.-K.C.); junyoung@cbnu.ac.kr (J.-Y.P.)

Abstract: Device guidelines for reducing power with punch-through current annealing in gate-all-around (GAA) FETs were investigated based on three-dimensional (3D) simulations. We studied and compared how different geometric dimensions and materials of GAA FETs impact heat management when down-scaling. In order to maximize power efficiency during electro-thermal annealing (ETA), applying gate module engineering was more suitable than engineering the isolation or source drain modules.

Keywords: annealing; dielectric; gate-all-around (GAA); hot-carrier injection (HCI); power consumption; punch-through; reliability; logic transistors

Citation: Kim, M.-K.; Choi, Y.-K.; Park, J.-Y. Power Reduction in Punch-Through Current-Based Electro-Thermal Annealing in Gate-All-Around FETs. *Micromachines* **2022**, *13*, 124. https://doi.org/10.3390/mi13010124

Academic Editor: Chengyuan Dong

Received: 15 December 2021
Accepted: 12 January 2022
Published: 13 January 2022

1. Introduction

MOSFETs have been aggressively scaled down to improve packing density and chip performance [1]. However, as semiconductor devices shrunk, several issues have arisen, such as short-channel effects (SCEs). SCEs give rise to an increase in the off-state current (I_{OFF}) and subthreshold swing (SS) and result in an increase in static power consumption ($P_{OFF} = V_{DD} \times I_{OFF}$) in the OFF-state. SCEs have been effectively suppressed by improving gate controllability not only with three-dimensional (3D) device structures such as FinFETs and gate-all-around (GAA) FETs, but also high-k gate dielectric and metal gate (HKMG) technology. In contrast to SCEs, improving device reliability during device minimization has become increasingly difficult. For example, recently, gate dielectric damage from hot-carrier injection (HCI), which is associated with the lateral drain electric field, has resurfaced as a matter of concern in semiconductor devices [2,3]. Typically, HCI increases both the threshold voltage (V_T) and SS, and hence results in unwanted V_T mismatching while also increasing I_{OFF} in circuitries. In addition, the HCI decreases both the ON-state current (I_{ON}) and lifetime, which affect chip speed and long-term usability, respectively [4,5].

To overcome the degradation of the gate dielectric, lightly doped drains (LDD) or forming gas annealing (FGA) have been more commonly used in mass production for decades [6,7]. However, it is difficult to realize long-term reliability longer than 10 years. Hence, electro-thermal annealing (ETA), which utilizes local heat generated by the device itself, has been introduced as a novel approach to cure the damaged gate dielectric [8].

It is possible that gate dielectric damage resulting from various stresses such as ionizing radiation, bias temperature instability, and HCI can be healed with the aid of ETA [8]. However, even though ETA can improve a device's reliability and lifetime, additional power consumption is inevitable, since ETA is performed by generating high-temperature Joule heating. To reduce power consumption an alternative is needed that would improve the power efficiency of ETA while enabling high-temperature generation.

In this work, the effects of geometric size and the material of the GAA FET were investigated to improve power efficiency during ETA. COMSOL simulation software was used to better understand the thermal dissipation and isolation characteristics during ETA. Temperature sensitivities were extracted and compared with respect to the gate module, including the gate electrode and gate spacer, source/drain module, and isolation layer.

2. Materials and Methods

Gate-all-around (GAA) FETs, fabricated on bulk wafer [9], as shown in Figure 1, were simulated as test specimens. The channel thickness (T_{Si}), channel width (W_{NW}), gate length (L_G), and gate height (H_G) were 20 nm, 20 nm, 60 nm, and 250 nm, respectively. The thickness of the gate hard mask (T_{HM}) and gate spacer (T_{SPC}), which are composed of SiO_2, were 50 nm and 30 nm, respectively.

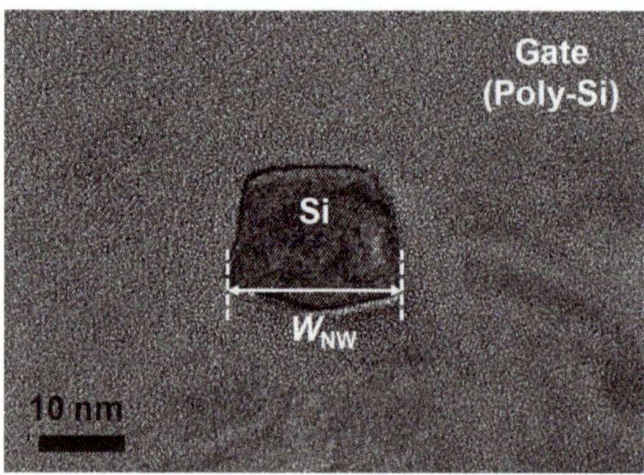

Figure 1. Transmission electron microscopy (TEM) image of the fabricated GAA FET.

Figure 2 shows a schematic of a GAA FET built on a bulk substrate for simulations. The Joule heating model in the heat transfer module of COMSOL was applied for 3D thermal profiling. During the simulation, the environment state and heat transfer coefficient (h) were assumed to be air and 10 W/m^{2}K, respectively. After that, punch-through current [10] was used for ETA instead of forward junction current [11] or gate-to-gate [12] current. Detailed device information used for the simulations are summarized in Table 1.

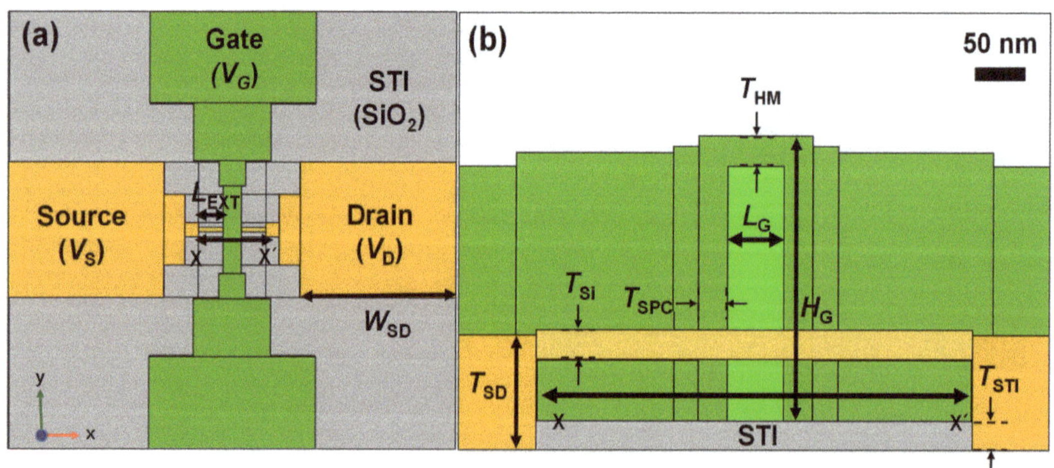

Figure 2. Schematic of the device used for simulations. (**a**) Top-view image of the GAA FET. (**b**) Cross-sectional image of the GAA FET cut along the x–x′ direction.

Table 1. Dimensional and material parameters for COMSOL simulations.

Geometry	Dimension	Material	Thermal Conductivity [W/m·K]
Gate length, L_G [nm]	60	Poly-Si	31.2
Gate height, H_G [nm]	300		
Gate hard mask thickness, T_{HM} [nm]	30	SiO_2	1
Gate spacer thickness, T_{SPC} [nm]	30		
Gate dielectric thickness, T_{GD} [nm]	5		
STI thickness, T_{STI} [nm]	70		
Source/drain pad thickness, T_{SD} [nm]	232	Si	149
Source/drain pad width, W_{SD} [nm]	1040		
Channel thickness, T_{Si} [nm]	20		
Channel width, W_{NW} [nm]	20		
Source/drain extension length, L_{EXT} [nm]	165		

3. Results and Discussion

Figure 3 shows the measured electrical I_D-V_G characteristics of the GAA FET. The DC characteristic was measured using a B1500A parameter analyzer at room temperature. After measurement of the initial state (e.g., initial state without stress), HCI stress at V_G = 2 V and V_D = 4 V was deliberately administered for 2 s. After the stress, degradation in the transconductance SS and V_T were observed at 227 mV/dec and 0.65 V, respectively. After that, bias conditions with V_G = 0.5 V and V_D = 6 V were applied for 100 µs to trigger a punch-through current-based ETA (Table 2). In fact, the current at the pinch-off is independent of V_G, and the V_G of 0.5 V was just referenced from our previous work [10]. After ETA, the aged-device characteristics with respect to SS and V_T recovered by 124 mV/dec and −0.05 V, respectively, compared to the initial state (Table 3). These facts show that both electrons were trapped in the gate dielectric, and physical damage at the SiO_2/Si interface was effectively cured by the punch-through current-based ETA.

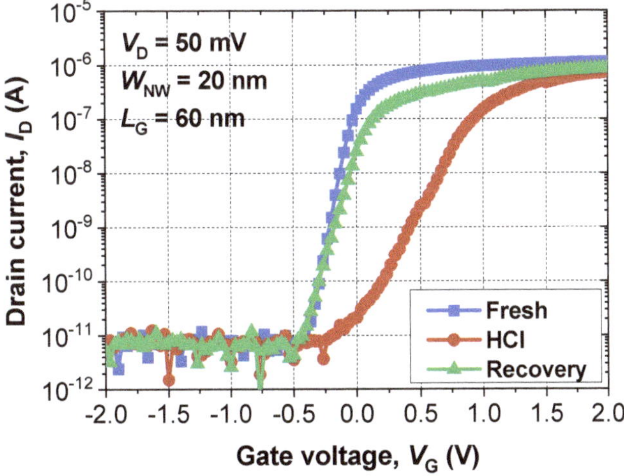

Figure 3. Measured I_D-V_G characteristic of the fabricated n-channel GAA FET.

Table 2. Bias conditions for punch-through current based ETA.

	Bias Condition
Gate voltage (V_G)	0.5 V
Source voltage (V_S)	0 V
Drain voltage (V_D)	6 V
Punch-through current (I_{Punch})	75 μA
Power consumption, ($P = V_D \times I_{Punch}$)	0.45 mW
Annealing time (t)	100 μs

Table 3. Extracted device parameters before HCI, after HCI and ETA.

	Initial State (Before HCI)	After HCI	After Punch-Through ETA
SS (mV/dec)	82 mV/dec	227 mV/dec	124 mV/dec
V_T (V)	−0.13 V	0.65 V	−0.05 V

Figure 4 shows a simulated heat distribution profile during ETA driven by the punch-through current in Figure 2. It shows that most of the heat during ETA was concentrated at the source/drain (S/D) extension where gate heat sink could not affect it. The extracted temperature at the S/D was symmetric [13]. However, considering the self-heating effect of semiconductor devices, the drain temperature was higher than that of the source region [14].

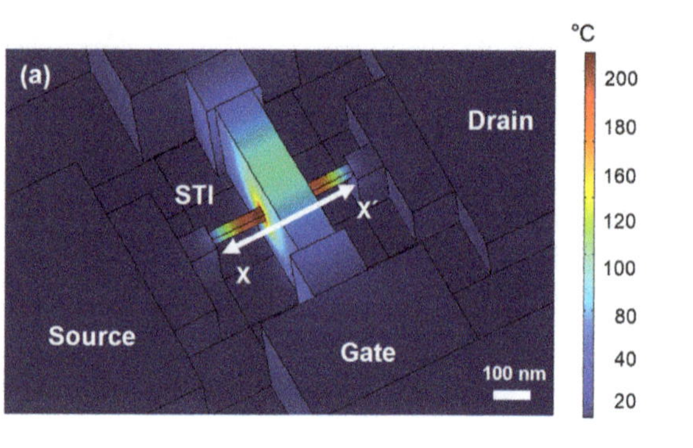

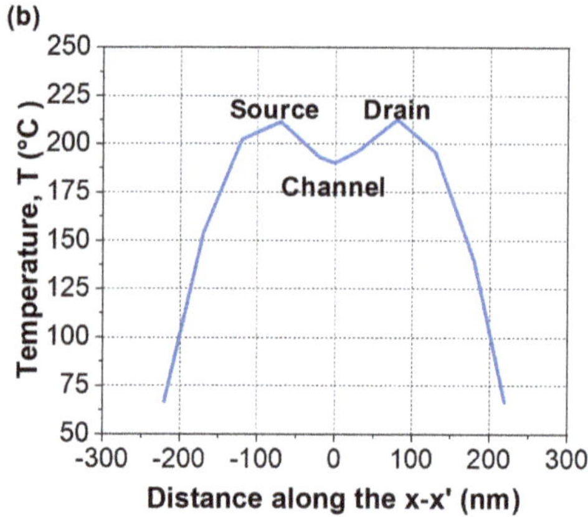

Figure 4. (**a**) Simulated heat distribution division profile during punch-through-based ETA under bias conditions with V_G = 0.5 V, $V_{S.}$ = 0 V, and V_D = 6 V. (**b**) Extracted temperature of the device along the x–x′ direction during ETA.

Figure 5 shows the extracted channel temperature ($T_{Channel}$) with respect to gate electrode scaling. All temperatures were extracted at the center of the silicon nanowire channel, i.e., $L_G/2$.

As the physical gate length and the height of the device were reduced, the temperature during ETA increased. Typically, the gate electrode acts as the heat sink during ETA. As the volume of the gate decreased, the temperature during ETA increased due to the reduced heat sink. The consistent high temperature generated during ETA under identical applied power consumption represented better power efficiency for gate dielectric curing. In this context, considering the extracted sensitivity of temperatures with respect to the gate length and the height, it would be more efficient to apply gate length scaling rather than the gate height. In addition, the gate module includes not only the gate electrode itself but also dielectric materials such as gate dielectric, gate spacer, and gate hard mask.

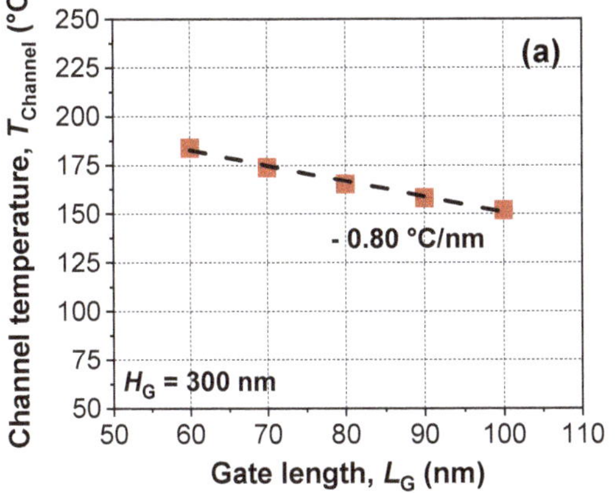

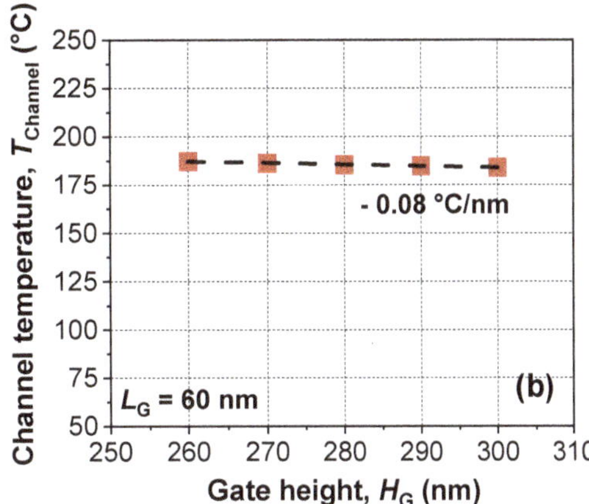

Figure 5. Extracted $T_{Channel}$ of devices with various (**a**) gate lengths and (**b**) gate heights under an identical power consumption of 0.45 mW. Dashed lines indicate linear fits of the experimental data.

Figure 6a shows the extracted $T_{Channel}$ with various thicknesses of gate dielectric composed of SiO_2. As the gate dielectric thickness (T_{GD}) increased, channel temperature increased under identical power consumption due to decreased heat dissipation through the gate electrode. However, considering the gate dielectric thickness was scaled down for better suppression of SCEs, this approach seems impractical for reducing power consumption. Alternatively, the material engineering shown in Figure 6b would be more efficient. As the thermal conductivity (κ) of the gate dielectric decreased, temperature sensitivity with applied power increased, due to increased thermal isolation, i.e., reduced heat dissipation with low κ.

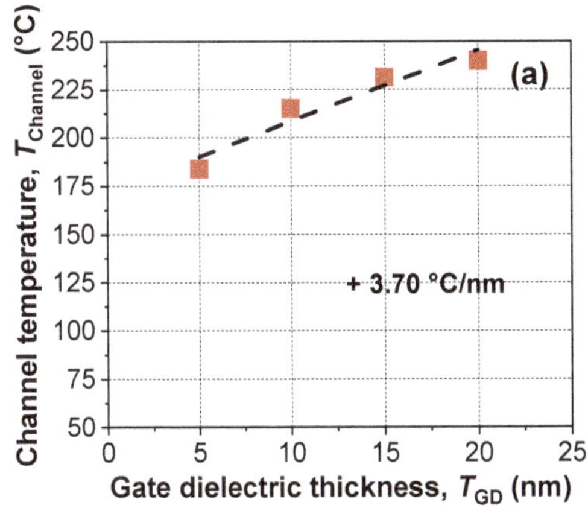

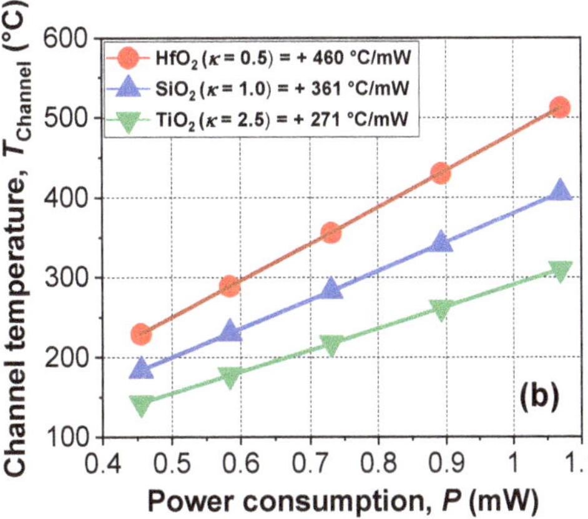

Figure 6. Extracted $T_{Channel}$ of devices with various (**a**) thicknesses and (**b**) materials of gate dielectric. Dashed lines indicate linear fits of the experimental data.

Figure 7a shows the extracted $T_{Channel}$ with various dielectric thicknesses of gate hard mask (T_{HM}) and gate spacer (T_{SPC}). The T_{HM} had a negligible effect on $T_{Channel}$ compared with the gate dielectric engineering in Figure 6. As the gate spacer increased, the temperature during ETA decreased due to the increased surface area of the gate spacer. Since convective cooling is performed through the air, a gate spacer with a small width and surface area would be more preferred to lower power consumption. Figure 7b shows the extracted channel temperature with various levels of thermal conductivity for the gate hard mask and the spacer. As the thermal conductivity of the dielectrics decreased, temperature sensitivity increased due to increased thermal isolation.

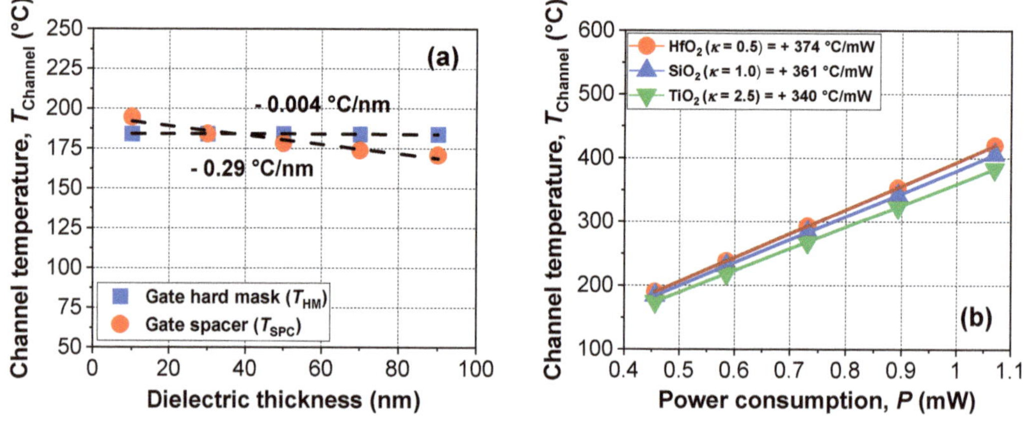

Figure 7. Extracted $T_{Channel}$ of devices with various (**a**) dielectric thicknesses and (**b**) materials of gate hard mask and gate spacer. Dashed lines indicate fitting of the symbols.

In contrast to the results in Figure 5 to Figure 7, which focused on the gate module, Figure 8 shows the device temperature with respect to modifications of the S/D module. However, even though the S/D extension showed the largest temperature sensitivity (Figure 8c), the sensitivity stemming from S/D was negligible. Moreover, considering the S/D extension (L_{EXT}) had been scaled down for better packing density, this approach seems impractical. In this context, reducing power consumption by engineering of the S/D module is not recommended.

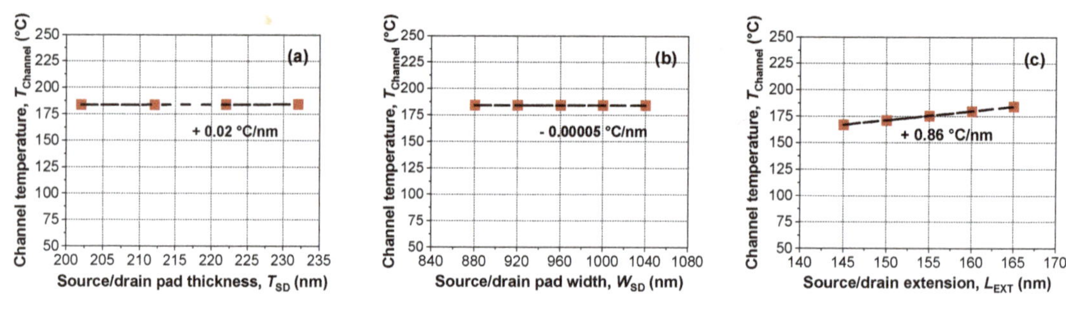

Figure 8. Extracted $T_{Channel}$ of devices with various (**a**) S/D pad thicknesses, (**b**) S/D pad widths, and (**b**,**c**) S/D extension lengths. Dashed lines indicate linear fits of the experimental data.

Figure 9a shows the extracted $T_{Channel}$ in the case of the isolation engineering by use of shallow trench isolation (STI) technology. As thickness T_{STI} increased, channel temperature could be increased due to the increased thermal isolation. However, the change was negligible because the channel was suspended from the STI. Figure 9b shows power efficiency with various buried dielectric materials. When a low thermal conductive material, e.g., HfO_2, is employed instead of SiO_2, the channel temperature could be increased under identical power consumption. Based on these results, our recommendation to maximize power efficiency is to apply low thermally conductive materials as an STI.

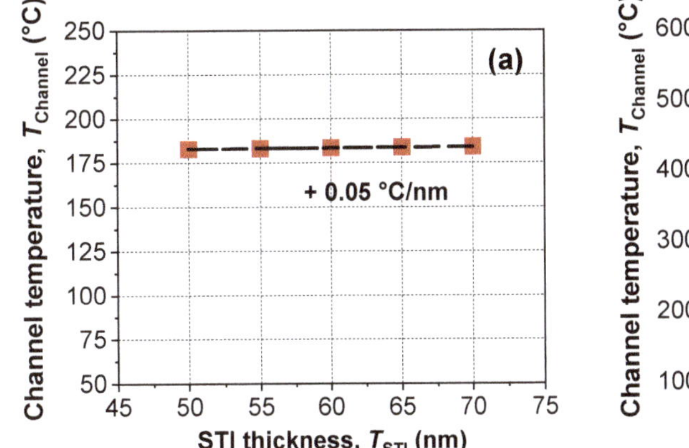

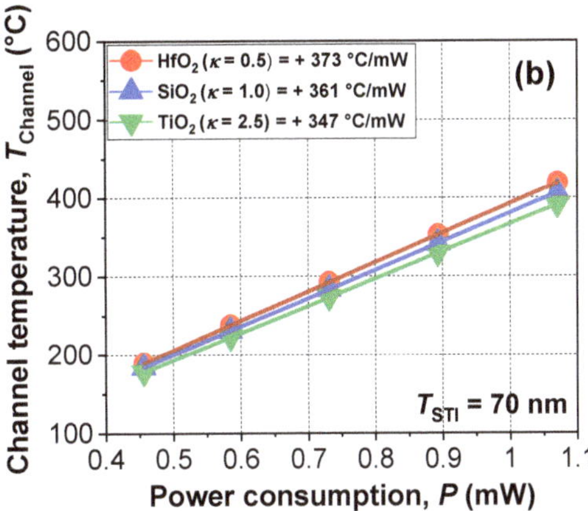

Figure 9. Extracted $T_{Channel}$ of devices with various (**a**) thickness and (**b**) materials of shallow trench isolation (STI). Dashed lines indicate linear fits of the experimental data.

Table 4 provides a summary of the temperature sensitivities for the different geometries and materials of the GAA FET. It can be concluded that the most significant design parameter for determining power efficiency is gate module engineering. As a result, the approach using gate module engineering would be more preferred to reducing power consumption for punch-through current-based ETA.

Table 4. Summary of temperature sensitivity according to dimensional and material engineering of the gate, S/D, and isolation module for the punch-through current-based local thermal annealing.

	Gate Module	S/D Module	Isolation
Minimum (°C/nm)	−0.80	0.00	+0.05
Maximum (°C/nm)	+3.70	+0.86	

4. Conclusions

Device guidelines for reducing the power of punch-through current annealing were investigated using 3D COMSOL simulations. Power management efficiency can be improved with dimensional and material engineering. The impacts of device scaling with respect to gate module, source/drain (S/D) module, and isolation, were compared in detail. The gate module engineering was found to be the most significant way to reduce power consumption. However, in contrast to the gate module, impacts of the S/D and shallow trench isolation were negligible.

Author Contributions: J.-Y.P. and Y.-K.C. conceived this project and designed all the experiments. M.-K.K. conducted all the simulations and wrote this paper. All authors have read and agreed to the published version of the manuscript.

Funding: This work was supported by the National Research Foundation (NRF) with a grant funded by the Korea government (MSIT) (No. 2020M3H2A1076786 and 2021R1F1A1049456).

Conflicts of Interest: The authors declare no conflict of interest.

References

1. Moroz, V.; Huang, J.; Arghavani, R. Transistor Design for 5 nm and Beyond: Slowing Down Electrons to Speed Up Transistors. In Proceedings of the 17th International Symposium on Quality Electronic Design (ISQED), Santa Clara, CA, USA, 15–16 March 2016; pp. 278–283. [CrossRef]
2. Ciou, F.-M.; Lin, J.-H.; Chen, P. H.; Chang, T.-C.; Chang, K.-C.; Hsu, J.-T.; Lin, Y.-S.; Jin, F.-Y.; Hung, W.-C.; Yeh, C.-H.; et al. Comparison of the Hot Carrier Degradation of N- and P-Type Fin Field-Effect Transistors in 14-nm Technology Nodes. *IEEE Electron Device Lett.* **2021**, *42*, 1420–1423. [CrossRef]

3. Lee, K.; Kaczer, B.; Kruv, A.; Gonzalez, M.; Degraeve, R.; Tyaginov, S.; Grill, A.; Wolf, I.D. Hot-Electron-Induced Punch-Through (HEIP) Effect in p-MOSFET Enhanced by Mechanical Stress. *IEEE Electron Device Lett.* **2021**, *42*, 1424–1427. [CrossRef]
4. Hu, C.; Simon, C.T.; Hsu, F.-C.; Ko, P.-K.; Chan, T.-Y.; Terrill, K.W. Hot-Electron-Induced MOSFET Degradation-Model, Monitor, and Improvement. *IEEE J. Solid State Circuit* **1985**, *20*, 295–305. [CrossRef]
5. Takeda, E.; Suzuki, N. An Empirical Model for Device Degradation Due to Hot-Carrier Injection. *IEEE Electron Dev. Lett.* **1983**, *4*, 111–113. [CrossRef]
6. Onishi, K.; Kang, C.S.; Choi, R.; Cho, H.-J.; Gopalan, S.; Nieh, R.E.; Krishnan, S.A.; Lee, J.C. Improvement of Surface Carrier Mobility of Hfo_2 Mosfets By High-Temperature Forming Gas Annealing. *IEEE Trans. Electron Devices* **2003**, *50*, 384–390. [CrossRef]
7. Jo, M.; Chang, M.; Park, H.; Hwang, H. Improvement of Hafnium Oxide/Silicon Oxide Gate Dielectric Stack Quality by High Pressure D_2O Post Deposition Annealing. *Jpn. J. Appl. Phys.* **2007**, *46*, L531–L533. [CrossRef]
8. Park, J.-Y.; Moon, D.-I.; Lee, G.-B.; Choi, Y.-K. Curing of Aged Gate Dielectric by the Self-Heating Effect in MOSFETs. *IEEE Trans. Electron Devices* **2020**, *67*, 777–788. [CrossRef]
9. Moon, D.-I.; Choi, S.-J.; Kim, C.-J.; Kim, J.-Y.; Lee, J.-S.; Oh, J.-S.; Lee, G.-S.; Park, Y.-C.; Hong, D.-W.; Lee, D.-W.; et al. Silicon Nanowire All-Around Gate Mosfets Built on a Bulk Substrate by All Plasma-Etching Routes. *IEEE Electron Device Lett.* **2011**, *32*, 452–454. [CrossRef]
10. Park, J.-Y.; Hur, J.; Choi, Y.-K. Demonstration of a Curable Nanowire FinFET Using Punchthrough Current to Repair Hot-Carrier Damage. *IEEE Electron Dev. Lett.* **2018**, *39*, 180–183. [CrossRef]
11. Lee, G.-B.; Kim, C.-K.; Park, J.-Y.; Bang, T.; Bae, H.; Kim, S.-Y.; Ryu, S.-W.; Choi, Y.-K. A Novel Technique for Curing Hot-Carrier-Induced Damage by Utilizing the Forward Current of the PN-Junction in a MOSFET. *IEEE Electron Dev. Lett.* **2017**, *38*, 1012–1014. [CrossRef]
12. Moon, D.-I.; Park, J.-Y.; Han, J.-W.; Jeon, G.-J.; Kim, J.-Y.; Moon, J.-B.; Seol, M.-L.; Kim, C.-K.; Lee, H.C.; Meyyappan, M.; et al. Sustainable Electronics for Nano-Spacecraft in Deep Space Missions. In Proceedings of the IEEE International Electron Devices Meeting (IEDM), San Francisco, CA, USA, 3–7 December 2016; pp. 3–7. [CrossRef]
13. Jeon, C.-H.; Park, J.-Y.; Seol, M.-L.; Moon, D.-I.; Hur, J.; Bae, H.; Jeon, S.-B.; Choi, Y.-K. Joule Heating to Enhance the Performance of a Gate-All-Around Silicon Nanowire Transistor. *IEEE Trans. Electron Devices* **2016**, *63*, 2288–2292. [CrossRef]
14. Pop, E.; Sinha, S.; Goodson, K.E. Heat Generation and Transport in Nanometer-Scale Transistors. *Proc. IEEE* **2006**, *94*, 1587–1601. [CrossRef]

MDPI

Article

An Artificial Synapse Based on CsPbI3 Thin Film

Jia-Ying Chen, Xin-Gui Tang *, Qiu-Xiang Liu, Yan-Ping Jiang, Wen-Min Zhong and Fang Luo

Guangzhou Higher Education Mega Center, School of Physics and Optoelectric Engineering, Guangdong University of Technology, Guangzhou 510006, China; 2112015044@mail2.gdut.edu.cn (J.-Y.C.); liuqx@gdut.edu.cn (Q.-X.L.); ypjiang@gdut.edu.cn (Y.-P.J.); zhongwen_min@163.com (W.-M.Z.); luofang_grsy9950@sina.com (F.L.)

* Correspondence: xgtang@gdut.edu.cn; Fax: +86-20-3932-2265

Abstract: With the data explosion in the intelligent era; the traditional von Neumann computing system is facing great challenges of storage and computing speed. Compared to the neural computing system, the traditional computing system has higher consumption and slower speed. However; the feature size of the chip is limited due to the end of Moore's Law. An artificial synapse based on halide perovskite $CsPbI_3$ was fabricated to address these problems. The $CsPbI_3$ thin film was obtained by a one-step spin-coating method, and the artificial synapse with the structure of Au/$CsPbI_3$/ITO exhibited learning and memory behavior similar to biological neurons. In addition, the synaptic plasticity was proven, including short-term synaptic plasticity (STSP) and long-term synaptic plasticity (LTSP). We also discuss the possibility of forming long-term memory in the device through changing input signals.

Keywords: artificial synapse; long-term synaptic plasticity; short-term synaptic plasticity; halide perovskite

Citation: Chen, J.-Y.; Tang, X.-G.; Liu, Q.-X.; Jiang, Y.-P.; Zhong, W.-M.; Luo, F. An Artificial Synapse Based on CsPbI3 Thin Film. *Micromachines* **2022**, *13*, 284. https://doi.org/10.3390/mi13020284

Academic Editors: Chengyuan Dong and Niall Tait

Received: 4 January 2022
Accepted: 8 February 2022
Published: 10 February 2022

1. Introduction

Integrated circuits have faced challenges. The large amounts of data limit the speed of calculation. The requirement of high-speed calculations causes large consumption. However, the end of Moore's Law means that the pursuit of high integration is no longer realistic. Therefore, it is necessary to design a high-speed, low-consumption computing system. In the brain, a nerve pulse arrives at each synapse about 10 times/s, on average. There are about 10^{15} synapses, so the brain accomplishes almost 10^{15} complex operations/s, with a cost of no more than 10^6 J each time [1,2]. As a new idea for solving problems, artificial synapses have attracted huge interest and have been widely studied.

Halide perovskites are advanced in artificial synapses, for which various defects in halide perovskites result in tunable charge capture capability. Tunable charge capture capability means alterable charge conductance and retention. In addition, halide perovskites have low halide ion migration activation energy, which is a convenient and optimized ionic semiconductor with excellent optical properties, providing conditions for the optical signal to become an effective modulation method for low-energy artificial synaptic devices [3]. Xu et al. reported an artificial synapse made from a bromine-containing organic halide perovskite (OHP), $MAPbBr_3$. This was the first time that OHP was applied to an artificial synapse [4]. Halide perovskite is also excellent in low-cost electric synapse, optical synapse and artificial synapse modulated by photoelectric combination [5–11]. CsPb $(I, Br, Cl)_3$, as a representation of all-inorganic halide perovskites (IHPs), has been continuously considered, owing to its excellent physical properties, such as longer carrier diffusion length and higher carrier mobility [12]. Compared with other materials used in artificial neural synapses, such as oxides and two-dimensional materials, the preparation conditions of $CsPbI_3$ are simpler and easier to synthesize [13–20]. In view of the excellent characteristics and neural morphology calculation of halide perovskites, it is necessary to further explore an artificial synaptic device with a simple preparation process and high performance. However, explorations of the application of IHPs to artificial synapses are remain rare.

Here, we report an artificial synapse based on $CsPbI_3$ thin film with the structure of Au/$CsPbI_3$/ITO. The synaptic plasticity in the Au/$CsPbI_3$/ITO artificial synapse is realized, including long-term and short-term plasticity, paired pulse facilitation, paired pulse inhibition, spike time-dependent plasticity, and spike number-dependent plasticity, which signifies the biological synaptic function is successfully simulated, including the possibility of forming long-term memory. The properties of neural synapses are simulated with a relatively simple structure and low-cost synthesized method (see Table S1).

2. Materials and Methods

We adopted one-step spin-coating to fabricate the $CsPbI_3$ thin film. We mixed CsI and PbI_2 with 1:1 mole proportion and added them to 10 mL DMF using magnetic stirring for 30 min at a temperature of 30 °C. A golden clear solution was obtained.

After standing for one day, we observed that the solution had no precipitation and used a dropper to take an appropriate amount of solution and drop it evenly on an ITO glass, with spin-coating speed of 500 rpm for 15 s and 2000 rpm for 30 s. A pale yellow film was obtained after annealing at 100 °C for 5 min. The ITO glass was soaked in absolute ethanol and underwent ultrasonic cleaning for 15 min.

The sample was plated with a 0.2 mm diameter Au point electrode to form a top–bottom electrode structure. This operation was completed by a small vacuum coating machine and a mask with a diameter of 0.2 mm at room temperature. The thickness and the surface morphology of the $CsPbI_3$ thin film was observed by field emission scanning electron microscopy (FESEM, Tescan, Brno, Czech Republic). Furthermore, the grazing-incidence X-ray diffraction (GIXRD, Bruker, Bremen, Germany) helped us complete the analysis of the phase structures of $CsPbI_3$ films. The biological synaptic characteristics and current-voltage (*I-V*) were measured by a Keithley 2400 instrument (Solon, OH, USA).

3. Results and Discussion

3.1. Structure Analysis

Peaks from the δ-phase can be seen from the X-ray diffraction (XRD) pattern [21]. As shown in Figure 1, the crystal phases of diffraction peaks at 21.8 degrees, 22.7 degrees, 26.6 degrees, 27.5 degrees, 30.3 degrees, and 37.6 degrees are (014), (023), (015), (032), (016), and (134), respectively. However, the PDF card (No. 18-0376) of the $CsPbI_3$ with yellow phase had diffraction peaks at 35.2 degrees, (133), perhaps indicating $CsPbI_3$ tends to change to yellow phase in the air.

The surface of the artificial synapse can be observed in Figure 2a; few particularly obvious pinholes are on the surface of the $CsPbI_3$ thin film, and the grains are relatively complete. As shown in Figure 2b, the thickness of the $CsPbI_3$ thin film is approximately 300 nm. As shown in Figure 3a, a pair of synapses, the basis of information transmission in the central nervous system, include the pre-neuron, the post-neuron, and the synaptic cleft. Cells maintain a resting potential unless stimulated, and once stimulated, they will produce an action potential, which is called a bioelectrical phenomenon. Inter-synaptic signals are transmitted by neurotransmitters, by which the action potential reaches the front of synapses; neurotransmitters are released because of Ca^{2+} entering pre-synaptic neurons to promote the combination of synaptic vesicles and the plasma membrane. Then, neurotransmitters combine with post-synaptic cell membrane receptors to change the post-synaptic potential and complete the transfer of action potential [22].

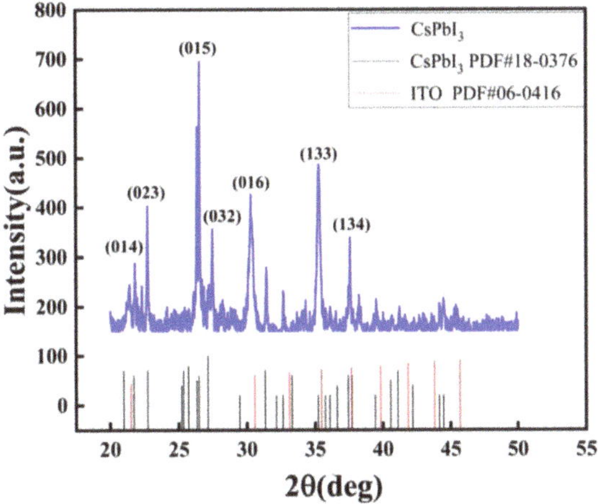

Figure 1. XRD patterns of the Au/$CsPbI_3$/ITO artificial synapse.

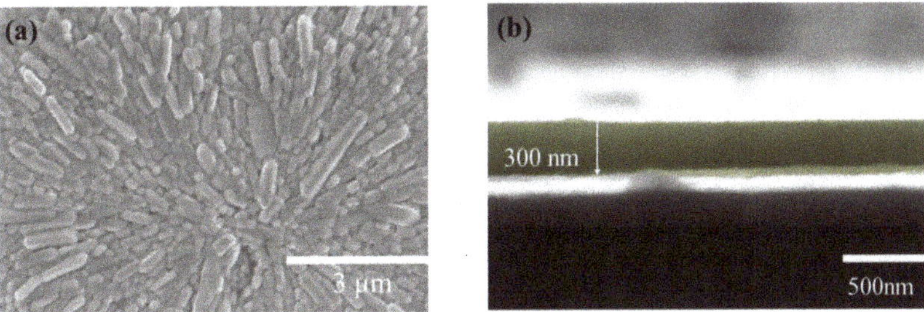

Figure 2. (**a**) The surficial SEM images of the Au/$CsPbI_3$/ITO artificial synapse; (**b**)the cross-sectional SEM images of the Au/$CsPbI_3$/ITO artificial synapse. The yellow part is $CsPbI_3$.

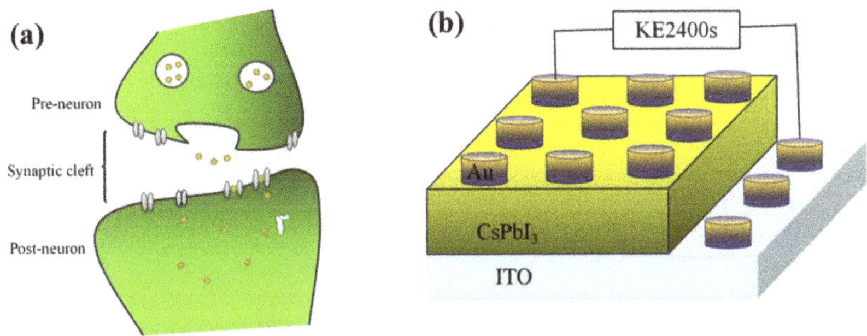

Figure 3. (**a**) A pair of synapses in biology; (**b**) the structure of the artificial synapse is Au/$CsPbI_3$/ITO. The top electron is Au, and the bottom electron is ITO.

Figure 3b shows that in the artificial synapse based on $CsPbI_3$, the top electrode is regarded as the pre-neuron, while the bottom electrode is regarded as the post-neuron. The insulator is similar to a synaptic cleft. Electrical signals are applied to the top and bottom electrodes to simulate the release process of neurotransmitters.

3.2. Memory Behaviors

Figure 4a shows the transition between high and low resistance states of the Au/$CsPbI_3$/ITO artificial synapse. When a negative voltage is applied, the iodine ions migrate to the bottom electrode, the top electrode and the bottom electrode are connected by the iodine vacancy conductive filaments, and the opposite voltage drives the filaments to decompose. Therefore, the device can switch between high and low resistance states. The formation and decomposition of conductive filaments make it possible for pulse modulation (voltage, spike-number, frequency) to induce the synaptic behavior of devices [23–28]. Figure 4b,c illustrate that the artificial synapse has learning and memory behavior similar to neural synapses. With a variety of stimuli, experience-dependent changes in synaptic weight occur, leading to the remodeling of neural circuits, which is called synaptic plasticity. The synaptic plasticity turns experience to memory, divided into short-term memory and long-term memory [29]. Controllable synaptic weight changes indicate that the device has basic learning and memory behaviors similar to synapses.

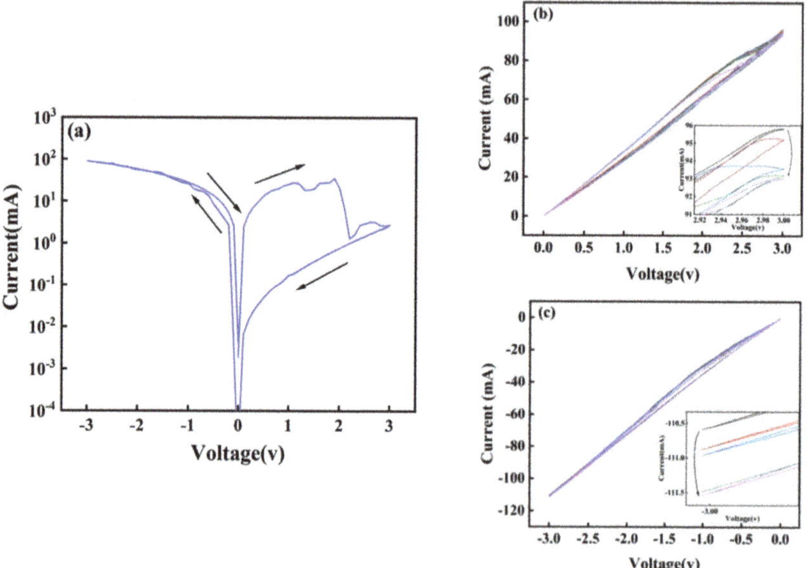

Figure 4. (**a**) The memory switching measurement of the Au/$CsPbI_3$/ITO artificial synapse. The memory behavior of the Au/$CsPbI_3$/ITO artificial synapse is measured by (**b**) five positive scanning voltage and (**c**) five negative scanning voltage.

3.3. The Synaptic Properties

In biology, synapses can be excitatory or inhibitory, specifically manifested as the increase or decrease of post-synaptic current (PSC). Figure 5a shows the signal transmission between two neurons. When the pre-spike arrives at the post-neuron, the state of post-neuron will change and be in a state of excitation or inhibition. Excitation and inhibition in a synapse depend on the change of the probability of action potential in post-synaptic cells, with the increase of action potential of pre-synaptic neurons [30]. Similarly, the conductance of the artificial synapse, a quantitative representation of the synaptic weight, can increase or decrease with the input voltage pulses.

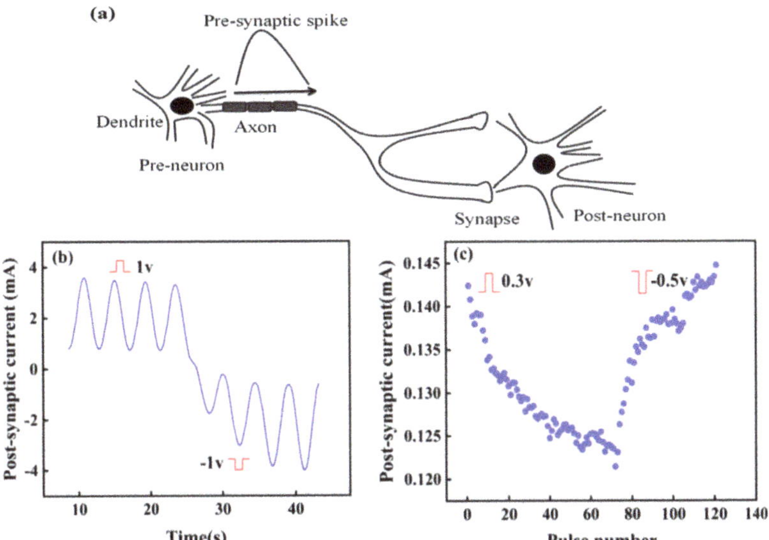

Figure 5. (**a**) Signal transmission between two neurons; (**b**) Response of current to time under continuous positive and negative pulses. (**c**) Potentiation and depression of current.

The artificial synapse has the characteristics of potential potentiation and depression by changing the polarity of input pulse. As shown in Figure 5b, stimulated by continuous positive voltage, the PSC of the artificial synapse decreases, while the PSC increases under continuous negative voltage. Figure 5c further reveals this phenomenon. When positive pulses are applied to the device, the PSC continues to decrease, indicating the synaptic device is in a state of excitation. However, the PSC continues to increase when pulses turn to negative ones, which demonstrates the artificial synapse is in the inhibition state. Excitation and inhibition can change the synaptic weight of the artificial synapse, further illustrating the synaptic plasticity in the artificial synapse, which can possibly contribute to memory formation.

3.4. The Synaptic Plasticity

Two manifestations are mentioned in the synaptic plasticity: one is short-term synaptic plasticity (STSP), and the other is long-term synaptic plasticity (LTSP). As the capacity of a pre-synaptic input to influence post-synaptic output, the effects of synaptic weight can be either enhancement or depression. In the STSP, enhancement and depression in synaptic weight only last for a short time. Compared to the STSP, in the LTSP, enhancement and depression can last for a longer time [31].

There are four states of potentiation and depression of the synaptic weight: short-term potentiation (STP), long-term potentiation (LTP), short-term depression (STD), and long-term depression (LTD). STSP includes STP and STD, while LTP and LTD are classified as LSTP. In the STP state, the synaptic weight rises temporally and then decreases rapidly to its original state. When repeated input pulses are applied, a permanent change occurs, suggesting the synaptic weight is in the LTP state. The STD state means the temporal decay of synaptic strength, but a permanent low value of synaptic weight represents the LTD state [32]. Shown in Figure 6a, the PSC increases after a long period of high-frequency stimulation, indicating the synaptic weight is enhanced. When $\Delta T = 0.3$ s lasts for 20 s, the PSC first increases and then tends to be flat, which indicates that repeated stimulation will stabilize the enhancement of synaptic weight and help to form long-term memory, while when $\Delta T = 1$ s lasts for 20 s, the PSC decreases, which indicates the artificial synapse is in a state of inhibition. However, Figure 6b shows that the short-term high-frequency stimulation will not enhance synaptic weight, but rather puts the synaptic weight in a LTD state.

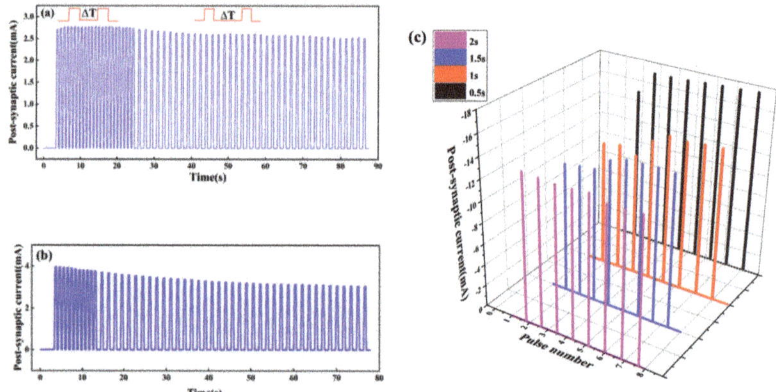

Figure 6. Response of current to voltage under two pulses with different time intervals (ΔT). The voltage of pulses is 0.5 v: (**a**) $\Delta T = 0.3$ s, lasting for 20 s, and $\Delta T = 1$ s, lasting for 60 s; (**b**) $\Delta T = 0.3$ s, lasting for 10 s, and $\Delta T = 1$ s, lasting for 60 s; (**c**) four sets of pulses with different ΔT are applied to the Au/$CsPbI_3$/ITO artificial synapse. The value of ΔT is 0.5, 1, 1.5, and 2 s, and the amplitude is −0.5 V.

When $\Delta T = 0.3$ s and lasts for 10 s, the PSC decreases, and the artificial synapse is inhibited. When $\Delta T = 1$ s, the PSC further decreases and finally tends to be flat, and the synaptic weight is in a LTD state.

The response of current to pulses at different ΔT is interesting. As shown in Figure 6c, four sets of pulses with different ΔT are applied to the Au/$CsPbI_3$/ITO synaptic device. When $\Delta T = 2$ s, the current decreases and the synaptic weight is inhibited, while when $\Delta T = 0.5$, 1, and 1.5 s, the current increases, indicating that synaptic weight is enhanced. In addition, compared to $\Delta T = 1$ s and $\Delta T = 1.5$ s, the current increases faster to the maximum and lasts longer when $\Delta T = 0.5$ s. It seems that pulses at a shorter time interval are conducive to the stable change of synaptic weight and the formation of long-term memory.

3.5. *The Spike-Number-Dependent of Au/$CsPbI_3$/ITO Artificial Synapse*

We also explored the spike-number-dependent (SNDP) of the artificial synapse, which is one of the synaptic properties [9]. It can be seen from Figure 7a,b, that when five different numbers of pulses are applied to the device, the SNDP index (In/I1) is obtained. With the increase of the number of pulses, the PSC gradually decreases, indicating artificial neurons are gradually inhibited. When fifty pulses are applied, the SNDP index reaches 80%, and artificial neurons are inhibited to the greatest extent compared to the first five pulses.

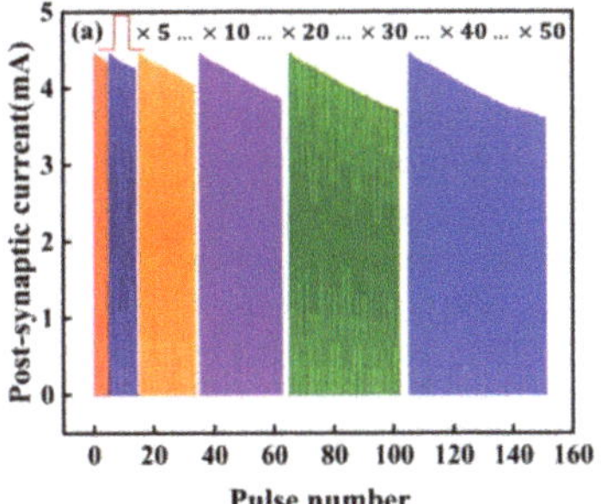

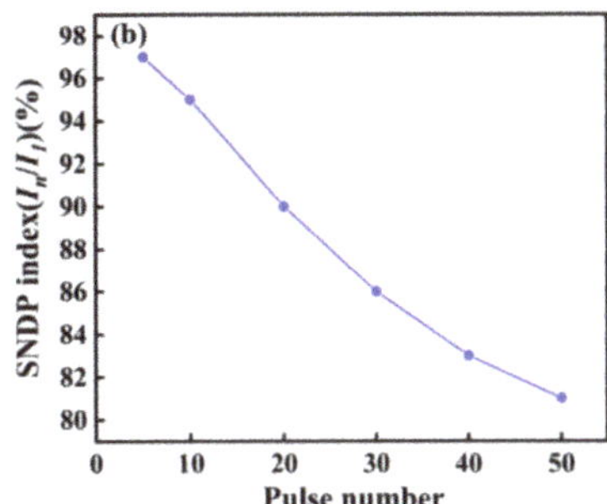

Figure 7. (**a**) SNDP; (**b**) SNDP index (I_n/I_1) of the Au/$CsPbI_3$/ITO synaptic artificial synapse. The number of pulses is 5,10, 20, 30, 40, and 50. (I_n/I_1) is the ratio of the amplitude of the PSC under the last pulse stimulation to the amplitude of PSC under the first pulse stimulation. The voltage of pulses is 0.5 V.

3.6. The Paired-Pulse Facilitation and Paired-Pulse Depression

As typical behaviors of STP and STD, the paired-pulse facilitation (PPF) and paired-pulse depression (PPD) are pivotal for both excitatory and inhibitory responses between adjacent synaptic connections [33]. Two pulses are applied to the Au/$CsPbI_3$/ITO artificial synapse. Compared to stimulation by the first pulse, the PSC increases after the second pulse if the device is in PPF state. The increase of post-synaptic current is inversely proportional to the time interval between two pulses. PPF and PPD are important in decoding the temporal information of visual and auditory signals, which reflects the emergence of recent spikes, so as to convert the temporal information into spatial data. In visual systems, via a pre-synaptic train of action potentials, PPF has dynamically controlled properties. It can endow different geniculate retina synapses, which makes synapses respond more abundantly on a time scale [34].

In order to study the STSP of the artificial synapse, we designed PPF and PPD experiments. As shown in Figure 8a,b, PPF and PPD were observed when a pair of pulses with an amplitude of 1 V were applied to the artificial synapse. In the PPF state, the PSC of the device increases after the last pulse, while in the PPD state, the PSC decreases. PPF and PPD illustrate the STSP in the artificial synapse, indicating the possibility of short-term memory formation.

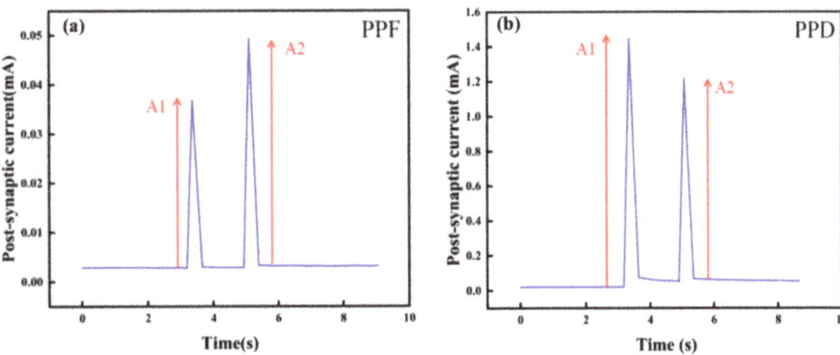

Figure 8. STSP of the Au/$CsPbI_3$/ITO artificial synapse. The amplitude of pulses is 1 V. A1 is the amplitude of the PSC after the first pulse. A2 is the amplitude of the PSC after the last pulse. (**a**) PPF state in the Au/$CsPbI_3$/ITO artificial synapse. (**b**) PPD state in the Au/$CsPbI_3$/ITO artificial synapse.

3.7. The Spike-Timing-Dependent Plasticity

Synapses play an important role in the transmission of signals between neurons and the memory and learning behaviors of the brain. Spike-timing-dependent plasticity (STDP), also called Hebbian learning, shows the fundamental function of synaptic plasticity in a biological synapse.

In the Hebb hypothesis, continuous and repeated stimulation will induce changes in the cells, such as some growth processes and metabolic changes. In the STDP state, pre-synaptic neurons continuously stimulate post-synaptic neurons, so the connection degree between neurons changes, which is an embodiment of LTSP. The concept explains cellular learning in the neural computing system. The asymmetric properties of STDP were found in cultures of hippocampal cells. The degree and direction of modification of synaptic weights is determined by the relative timing of pre-synaptic spikes and post-synaptic spikes. When a pre-synaptic spike arrives at time t_{pre} before or after the post-synaptic spike time t_{post}, the synaptic weight changes, and LTP or LTD behavior is induced [35].

In biology, the STDP learning rule is a typical LTSP. As one of the basic learning rules for simulating synaptic functions, the synaptic weight can be enhanced/inhibited by adjusting the spike time or sequence through STDP [36]. For the sake of the STDP function realization, pulses with an amplitude of 0.3 V and −0.5V are imposed on the artificial synapse as the pre-synaptic spikes and post-synaptic spikes. Figure 9 displays the STDP in the artificial synapse. The relative change of synaptic weight (ΔW) changes with the change of relative time (Δt). The artificial synapse is in the LTP state is at $\Delta t < 0$. However, the artificial synapse turns to LTD at $\Delta t > 0$.

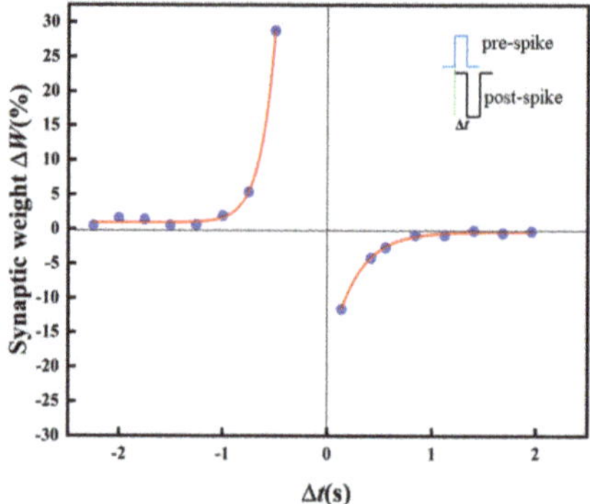

Figure 9. STDP of the Au/$CsPbI_3$/ITO artificial synapse. The relative change of the synaptic weight (ΔW) versus the relative spike timing (Δt); $\Delta t = t_{post} - t_{pre}$. The blue dots are the measured data, and the red solid line is the result of fitting by the Origin 2021.

In the case of pre-synaptic spikes or post-synaptic spikes stimulated earlier than post-synaptic spikes or pre-synaptic spikes, long-term potentiation or inhibition is observed, and the absolute value of ΔW increases exponentially with decreasing Δt. The STDP in biological synapses is expressed as follows, including the Symmetric Hebbian and antisymmetric Hebbian learning rules [37].

$$\Delta W = A \exp(-\Delta t^2/\tau^2) + \Delta W_0 \tag{1}$$

$$\Delta W = A \exp(-\Delta t/\tau) + \Delta W_0 \tag{2}$$

A and τ are the scaling factor and time constant, respectively, and ΔW is the relative change of the synaptic weight ($\Delta W = (W_{\mathrm{post}} - W_{\mathrm{pre}})/W_{\mathrm{pre}}$). The experimental data of ΔW and τ could be well fitted with the exponential decay function (Equation (2)) in the antisymmetric Hebbian learning rules, where A_1 is -19.4, τ_1 is $0.26(\Delta t > 0)$, A_2 is 1078.7, and τ_2 is -0.14 ($\Delta t < 0$).

4. Conclusions

We designed an artificial synapse based on halide perovskite. We successfully fabricated the CsPbI3 thin film and adopted a structure of Au/$CsPbI_3$/ITO. The Au/$CsPbI_3$/ITO artificial synapse exhibited depression/potentiation after positive/negative pulses, which indicated a performance similar to that of a neural synapse in biology. In addition, the Au/$CsPbI_3$/ITO artificial synapse exhibited favorable synaptic plasticity, including STSP and LTSP. STSP is divided into STP and STD. The device displayed a typical behavior of STP (PPF) and another typical behavior of STD (PPD). The phenomenon of STDP indicates LTP and LTD in the device, which is a typical behavior of LTSP. We explored the possibility of long-term memory formation through applying pulses with different time intervals and different pulse durations. The results indicate that the high-density durable stimulation is conducive to long-term memory formation. As one of the synaptic properties, the device also reflects the SNDP and the SNDP index, showing that the number of pulses causes a change of the PSC.

Supplementary Materials: The following supporting information can be downloaded at: https://www.mdpi.com/article/10.3390/mi13020284/s1, Table S1: Structure, Synthesized method and Synaptic properties of artificial neural synapses based on oxide and 2D materials.

Author Contributions: J.-Y.C. conducted the experiment. X.-G.T. conceived the idea. J.-Y.C., W.-M.Z. and F.L. contributed to some characterizations of perovskite films. X.-G.T., Q.-X.L. and Y.-P.J. supervised the project. J.-Y.C. and X.-G.T. wrote the paper. All authors have read and agreed to the published version of the manuscript.

Funding: This research was funded by "National Natural Science Foundation of China" (Grant Nos 11574057, and 12172093) and the "Guangdong Basic and Applied Basic Research Foundation" (Grant No.2021A1515012607).

Data Availability Statement: The data that support the funding of this study are available from the corresponding author upon reasonable request.

Conflicts of Interest: The authors declare no conflict of interest.

References

1. Ho, V.M.; Lee, J.-A.; Martin, K.C. The Cell Biology of Synaptic Plasticity. *Science* **2011**, *334*, 623–628. [CrossRef]
2. Mead, C. Neuromorphic electronic systems. *Proc. IEEE* **1990**, *78*, 1629–1636. [CrossRef]
3. Chen, S.; Huang, J. Recent Advances in Synaptic Devices Based on Halide Perovskite. *ACS Appl. Electron. Mater.* **2020**, *2*, 1815–1825. [CrossRef]
4. Xu, W.; Cho, H.; Kim, Y.-H.; Wolf, C.; Park, C.-G.; Lee, T.-W. Organometal Halide Perovskite Artificial Synapses. *Adv. Mater.* **2016**, *28*, 5916–5922. [CrossRef] [PubMed]
5. Wang, K.; Dai, S.; Zhao, Y.; Wang, Y.; Liu, C.; Huang, J. Light-Stimulated Synaptic Transistors Fabricated by a Facile Solution Process Based on Inorganic Perovskite Quantum Dots and Organic Semiconductors. *Small* **2019**, *15*, e1900010. [CrossRef]
6. Zhu, X.; Lu, W.D. Optogenetics-Inspired Tunable Synaptic Functions in Memristors. *ACS Nano* **2018**, *12*, 1242–1249. [CrossRef]
7. Ham, S.; Choi, S.; Cho, H.; Na, S.-I.; Wang, G. Photonic organolead halide perovskite artificial synapse capable of accelerated learning at low power inspired by dopamine-facilitated synaptic activity. *Adv. Funct. Mater.* **2019**, *29*, 1806646. [CrossRef]
8. Wang, Y.; Lv, Z.; Chen, J.; Wang, Z.; Zhou, Y.; Zhou, L.; Chen, X.; Han, S.T. Photonic synapses based on inorganic perovskite quantum dots for neuromorphic computing. *Adv. Mater.* **2018**, *30*, 1802883. [CrossRef]
9. Kim, S.I.; Lee, Y.; Park, M.H.; Go, G.T.; Kim, Y.H.; Xu, W.; Lee, H.D.; Kim, H.; Seo, D.G.; Lee, W.; et al. Dimensionality dependent plasticity in halide perovskite artificial synapses for neuromorphic computing. *Adv. Electron. Mater.* **2019**, *5*, 1900008. [CrossRef]
10. Kim, S.G.; Van Le, Q.; Han, J.S.; Kim, H.; Choi, M.J.; Lee, S.A.; Kim, T.L.; Kim, S.B.; Kim, S.Y.; Jang, H.W. Dual-phase all-inorganic cesium halide perovskites for conducting-bridge memory-based artificial synapses. *Adv. Funct. Mater.* **2019**, *29*, 1906686. [CrossRef]
11. Guan, X.; Wang, Y.; Lin, C. H.; Hu, L.; Ge, S.; Wan, T.; Younis, A.; Li, F.; Cui, Y.; Qi, D. C.; et al. A monolithic artificial iconic memory based on highly stable perovskite-metal multilayers. *Appl. Phys. Rev.* **2020**, *7*, 031401. [CrossRef]

12. Saleem, M.I.; Yang, S.; Batool, A.; Sulaman, M.; Veeramalai, C.P.; Jiang, Y.; Tang, Y.; Cui, Y.; Tang, L.; Zou, B. CsPbI3 nanorods as the interfacial layer for high-performance, all-solution-processed self-powered photodetectors. *J. Mater. Sci. Technol.* **2021**, *75*, 196–204. [CrossRef]
13. Seo, S.; Lee, J.-J.; Lee, H.-J.; Lee, H.W.; Oh, S.; Heo, K.; Park, J.-H. Recent Progress in Artificial Synapses Based on Two-Dimensional van der Waals Materials for Brain-Inspired Computing. *ACS Appl. Electron. Mater.* **2020**, *2*, 371–388. [CrossRef]
14. Luo, Z.-D.; Xia, X.; Yang, M.-M.; Wilson, N.R.; Gruverman, A.; Alexe, M. Artificial Optoelectronic Synapses Based on Ferroelectric Field-Effect Enabled 2D Transition Metal Dichalcogenide Memristive Transistors. *ACS Nano* **2019**, *14*, 746–754. [CrossRef] [PubMed]
15. Huh, W.; Lee, D.; Lee, C. Memristors Based on 2D Materials as an Artificial Synapse for Neuromorphic Electronics. *Adv. Mater.* **2020**, *32*, 2002092. [CrossRef]
16. Sun, L.; Yu, H.; Wang, D.; Jiang, J.; Kim, D.; Kim, H.; Zheng, S.; Zhao, M.; Ge, Q.; Yang, H. Selective Growth of Monolayer Semiconductors for Diverse Synaptic Junctions. *2D Mater.* **2018**, *6*, 015029. [CrossRef]
17. Wang, S.; Dang, B.; Sun, J.; Zhao, M.; Yang, M.; Ma, X.; Wang, H.; Hao, Y. Physically Transient W/ZnO/MgO/W Schottky Diode for Rectifying and Artificial Synapse. *IEEE Electron Device Lett.* **2020**, *41*, 844–847. [CrossRef]
18. Lu, K.; Li, X.; Sun, Q.; Pang, X.; Chen, J.; Minari, T.; Liu, X.; Song, Y. Solution-processed electronics for artificial synapses. *Mater. Horiz.* **2020**, *8*, 447–470. [CrossRef]
19. Yoon, S.J.; Moon, S.E.; Yoon, S.M. Implementation of an electrically modifiable artificial synapse based on ferroelectric field-effect transistors using Al-doped HfO2 thin films. *Nanoscale* **2020**, *12*, 13421–13430. [CrossRef]
20. Mohta, N.; Mech, R.K.; Sanjay, S.; Muralidharan, R.; Nath, D.N. Artificial synapse based on back-gated MoS2 field-effect transistor with high-k Ta2O5 dielectrics. *Phys. Status Solidi A* **2020**, *217*, 2000254. [CrossRef]
21. Ge, J.; Ma, Z.; Chen, W.; Cao, X.; Yan, J.; Fang, H.; Qin, J.; Liu, Z.; Pan, S. Solution-processed inorganic δ-phase CsPbI3 electronic synapses with short- and long-term plasticity in a crossbar array structure. *Nanoscale* **2020**, *12*, 13558–13566. [CrossRef]
22. Sun, K.; Chen, J.; Yan, X. The Future of Memristors: Materials Engineering and Neural Networks. *Adv. Funct. Mater.* **2021**, *31*, 2006773. [CrossRef]
23. Bousoulas, P.; Papakonstantinopoulos, C.; Kitsios, S.; Moustakas, K.; Sirakoulis, G.C.; Tsoukalas, D. Emulating Artificial Synaptic Plasticity Characteristics from SiO2-Based Conductive Bridge Memories with Pt Nanoparticles. *Microm Achines* **2021**, *12*, 306. [CrossRef]
24. Cho, H.; Kim, S. Emulation of Biological Synapse Characteristics from Cu/AlN/TiN Conductive Bridge Random Access Memory. *Nanomaterials* **2020**, *10*, 1709. [CrossRef] [PubMed]
25. Ge, S.; Wang, Y.; Xiang, Z.; Cui, Y. Reset Voltage-Dependent Multilevel Resistive Switching Behavior in CsPb1xBixI3 Perovskite-Based Memory Device. *ACS Appl. Mater. Interfaces* **2018**, *10*, 24620–24626. [CrossRef] [PubMed]
26. Liu, Z.; Cheng, P.; Li, Y.; Kang, R.; Zhang, Z.; Zuo, Z.; Zhao, J. High Temperature CsPbBrxI3x Memristors Based on Hybrid Electrical and Optical Resistive Switching Effects. *ACS Appl. Mater. Interfaces* **2021**, *13*, 58885–58897. [CrossRef] [PubMed]
27. Shen, Z.; Zhao, C.; Qi, Y.; Xu, W.; Liu, Y.; Mitrovic, I.Z.; Yang, L.; Zhao, C. Advances of RRAM Devices: Resistive Switching Mechanisms, Materials and Bionic Synaptic Application. *Nanomaterials* **2020**, *10*, 1437. [CrossRef]
28. Zhong, W.-M.; Liu, Q.-X.; Jiang, Y.-P.; Deng, M.-L.; Li, W.-P.; Tang, X.-G. Ultra-high dielectric tuning performance and double-set resistive switching effect achieved on the Bi2NiMnO6 thin film prepared by solgel method. *J. Colloid Interface Sci.* **2021**, *606*, 913–919. [CrossRef] [PubMed]
29. Takeuchi, T.; Duszkiewicz, A.; Morris, R.G.M. The synaptic plasticity and memory hypothesis: Encoding, storage and persistence. *Philos. Trans. R. Soc. B Biol. Sci.* **2014**, *369*, 20130288. [CrossRef]
30. Zhu, M.; He, T.; Lee, C. Technologies toward next generation human machine interfaces: From machine learning enhanced tactile sensing to neuromorphic sensory systems. *Appl. Phys. Rev.* **2020**, *7*, 031305. [CrossRef]
31. Tang, J.; Yuan, F.; Shen, X.; Wang, Z.; Rao, M.; He, Y.; Sun, Y.; Li, X.; Zhang, W.; Li, Y.; et al. Bridging Biological and Artificial Neural Networks with Emerging Neuromorphic Devices: Fundamentals, Progress, and Challenges. *Adv. Mater.* **2019**, *31*, e1902761. [CrossRef] [PubMed]
32. Gil Kim, S.; Han, J.S.; Kim, H.; Kim, S.Y.; Jang, H.W. Recent Advances in Memristive Materials for Artificial Synapses. *Adv. Mater. Technol.* **2018**, *3*, 1800457. [CrossRef]
33. Min, S.-Y.; Cho, W.-J. Memristive Switching Characteristics in Biomaterial Chitosan-Based Solid Polymer Electrolyte for Artificial Synapse. *Int. J. Mol. Sci.* **2021**, *22*, 773. [CrossRef]
34. Bian, H.; Goh, Y.Y.; Liu, Y.; Ling, H.; Xie, L.; Liu, X. Stimuli-Responsive Memristive Materials for Artificial Synapses and Neuromorphic Computing. *Adv. Mater.* **2021**, *33*, 2006469. [CrossRef] [PubMed]
35. Park, Y.; Kim, M.-K.; Lee, J.-S. Emerging memory devices for artificial synapses. *J. Mater. Chem. C* **2020**, *8*, 9163–9183. [CrossRef]
36. Lin, Y.; Zeng, T.; Xu, H.; Wang, Z.; Zhao, X.; Liu, W.; Ma, J.; Liu, Y. Transferable and flexible artificial memristive synapse based on WOx schottky junction on arbitrary substrates. *Adv. Electron. Mater.* **2018**, *4*, 1800373. [CrossRef]
37. Lao, J.; Xu, W.; Jiang, C.; Zhong, N.; Tian, B.; Lin, H.; Luo, C.; Travas-Sejdic, J.; Peng, H.; Duan, C.-G. An air-stable artificial synapse based on a lead-free double perovskite Cs2AgBiBr6 film for neuromorphic computing. *J. Mater. Chem. C* **2021**, *9*, 5706–5712. [CrossRef]

micromachines

Article

A Reconfigurable Surface-Plasmon-Based Filter/Sensor Using D-Shaped Photonic Crystal Fiber

S. Selvendran [1,*], J. Divya [1], A. Sivanantha Raja [2], A. Sivasubramanian [1] and Srikanth Itapu [3]

[1] School of Electronics Engineering (SENSE), Vellore Institute of Technology, Chennai 600127, Tamil Nadu, India; divyaselva0605@gmail.com (J.D.); sivasubramanian.a@vit.ac.in (A.S.)

[2] Department of ECE, Alagappa Chettiar Government College of Engineering and Technology, Karaikudi 630003, Tamil Nadu, India; sivanantharaja@yahoo.com

[3] Department of ECE, Alliance College of Engineering and Design, Alliance University, Bengaluru 562106, Karnataka, India; srikanth.itapu@alliance.edu.in

* Correspondence: selvendrans@aol.com or selvendran.s@vit.ac.in

Abstract: A reconfigurable surface-plasmon-based filter/sensor using D-shaped photonic crystal fiber is proposed. Initially a D-shaped PCF is designed and optimized to realize the highly birefringence and by ensuring the single polarization filter. A tiny layer of silver is placed on the flat surface of the D-shaped fiber with a small half-circular opening to activate the plasmon modes. By the surface plasmon effect a maximum confinement loss of about 713 dB/cm is realized at the operating wavelength of 1.98 μm in X-polarized mode. At this wavelength the proposed fiber only allows Y-polarization and filters the X-polarization using surface plasmon resonance. It is also exhibiting maximum confinement loss of about 426 dB/cm at wavelength 1.92 μm wavelength for Y-polarization. At this 1.92 μm wavelength the proposed structure attenuated the Y-polarization completely and allowed X-polarization alone. The proposed PCF polarization filter can be extended as a sensor by adding an analyte outside this filter structure. The proposed sensor can detect even a small refractive index (RI) variation of analytes ranging from 1.34–1.37. This sensor provides the maximum sensitivity of about 5000 nm/RIU; it enables this sensor to be ideally suited for various biosensing and industrial applications.

Keywords: polarization filter; photonic crystal fiber; surface plasmon resonance; plasmonic sensor; silver

Citation: Selvendran, S.; Divya, J.; Sivanantha Raja, A.; Sivasubramanian, A.; Itapu, S. A Reconfigurable Surface-Plasmon-Based Filter/Sensor Using D-Shaped Photonic Crystal Fiber. *Micromachines* **2022**, *13*, 917. https://doi.org/10.3390/mi13060917

Academic Editors: Chengyuan Dong and Jaeyoun Kim

Received: 17 April 2022
Accepted: 30 May 2022
Published: 9 June 2022

1. Introduction

Photonic crystal fibers (PCF) are a new type of fiber that has a flexible structure and various unique features, such as high nonlinearity [1], engineered dispersion profile over the region of interest [2,3], high birefringence [4], endless single mode, low transmission loss [2–5] and easy filling of materials in fiber [6–8]. PCF has been extended to plasmonic devices by filling them with various metals, such as gold [9–13], silver [14,15], copper [16] and titanium [17]. Surface plasmon (SP) develops at the metal/dielectric interface when plasmonic materials are exposed to electromagnetic waves due to the collective oscillations of conduction electrons [18]. At the phase-matching condition, the core-guided mode transfers maximum energy to the plasmon mode. As a result of coupling, core mode energy decreases, and confinement loss increases. This phenomenon is known as surface plasmon resonance (SPR) [19].

PCF-based plasmonic devices have gained popularity because of their small size, versatility and controllability. Sensors [20], polarization filters [21], multiplexers–demultiplexers [22] and polarization splitters [23] are few examples of PCF-based plasmonic devices applications. In communication systems, a polarization filter is an essential element. Metal-coated or metal-filled plasmonic polarization filters have gained a lot of interest in recent years. When the core-guided mode is phase matched with the plasmon mode at a specific wavelength, SPR occurs. It will also absorb the power of one polarized mode while allowing the other to pass through. As a result, a PCF-based polarization filter can filter different polarizations. The SPR is present in only a few metals; these noble metals, such as gold and silver, are widely investigated in SPR due to their high stability.

In recent years, many researchers have proposed novel PCF-based polarization filters. In 2017, Ying et al. reported a silver-filled tunable single-polarization filter. Their reported confinement losses for X-polarization were 371 dB/cm and 252 dB/cm at the operating wavelengths of 1310 nm and 1550 nm, respectively, whereas Y-polarization mode exhibited very low confinement losses of about 14 dB/cm and 10 dB/cm, respectively. For a propagation distance of 1 mm, a bandwidth of 179 and 71 nm was obtained at 1310 and 1550 nm wavelengths [24]. Manish Sharma, et al. (2019) reported an RI sensor using an annular core photonic crystal fiber. This work exhibits high sensitivity over a large dynamic range. In particular, in the RI range from 1.31 to 1.39, this proposed fiber displays more propagation loss to sense the biochemicals, and its recorded sensitivity is 2.65×10^4 (dB/m)/RIU at 1550 nm and 7.83×10^3 (dB/m/) RIU at 980 nm [25]. Xue et al. published a paper in 2018 that described a novel offset core filled with a gold layer photonic crystal fiber filter. At a wavelength of 1.55 μm, the confinement loss of a Y-polarized light was 657 dB/cm, while the X-polarized light was comparatively low. At a wavelength of 1.55 μm, the cross-talk of 56.2 dB with a bandwidth of 100 nm was obtained for a fiber length of 1 μm [26]. In that year, Yang et al. developed a high-birefringence PCF filter with silver layers that was selectively coated. At a 1.31 μm operating wavelength, a confinement loss of 500 dB/cm was obtained, and a full-width half-maximum was 23 nm [27]. Liu et al. suggested a single-polarization bimetal-coated (Au/Ag) PCF filter with liquid-filled air holes in 2019. Confinement losses for the Y-polarized mode were 544.3 dB/cm and 147.3 dB/cm for a wavelength of 1.31 μm and 1.55 μm, respectively. In the same wavelength, the X-polarized losses were 12.3 dB/cm and 24.0 dB/cm. At a wavelength of 1.31 μm and 1.55 μm, the cross-talk of 462 dB and 107.1 dB was achieved, respectively, using 224 nm and 504 nm bandwidths [28]. Yan et al. published a new gold-plated PCF polarization filter in 2020. There were two big air holes in the filter, which was selectively coated with gold after being filled with water. At 1310 nm, the confinement loss in the Y-polarized mode was 1209.57 dB/cm, but the confinement loss in the X-polarized mode was almost non-existent [29]. Lavanya et al. reported a combined silver–graphene layered pentagonal PCF filter in the year 2021. The filter renders an X-polarized mode with losses of 0.12 dB/cm and 0.59 dB/cm at wavelengths of 1.31 μm and 1.55 μm, respectively. It also rejects a Y-polarized mode with a loss of 361.26 dB/cm and 1508.37 dB/cm, respectively. At 1.31 μm and 1.55 μm, respectively, the cross-talk of 216.68 dB and 904.67 dB was obtained for fiber lengths of 300 μm. [30]. Coating a metal in the inner part of the fiber is complex and hard to fabricate, but it is used in the majority of the filters. A D-shaped plasmonic fiber has been proposed to tackle this problem, with the top circular portion of the fiber removed and the top surface polished. In 2018, Ying et al. reported a D-shaped PCF single-polarization filter. A micro-opening had formed at the top of the PCF, and a gold layer was coated on it. At 1.55 μm, the Y-polarized mode's confinement loss reached 376.31 dB/cm, while the X-polarized mode's loss was 0.17 dB/cm. For a fiber length of 1 mm, a bandwidth of 480 nm was attained with a crosstalk of 326.7 dB/cm [31]. In 2019, Almewafy et al. proposed an improved D-shaped filter that deposits gold on both sides. At an operating wavelength of 1.3 μm and 1.52 μm, the crosstalk was 46.4 and 41.2 dB, respectively, while the bandwidth was 45 nm and 110 nm for a fiber length of 23 μm [32]. In 2020, Shima et al. proposed a bimetallic coated D-shaped PCF-based single-polarization filter. A Y-polarized mode's confinement loss reached 950.68 dB/cm at a wavelength of 1.55 μm. The bandwidth was 380 nm, with an extinction ratio of 806.52 dB [33].

SPR technique is also widely used in sensing applications because of its accuracy and extreme sensing performance. At the phase-matching condition, maximal energy transferred from the fundamental mode to the plasmon mode causes a shift in resonance wavelength. The resonance wavelength shift method is used in the SPR-based sensor. It can detect slight variation in the refractive index of the analyte. The external sensing technique is used because the filling and removal of analytes within the structure are complicated [34]. Rifat et al. proposed a PCF-based SPR sensor. For better detection, silver was utilized as a plasmonic material and was coated with a graphene layer. The external sensing technique was used to detect the refractive index of the analyte and obtained the sensitivity of 3000 nm/RIU for the sensing range of 1.46–1.49 [35]. The surface-plasmon-based sensors using silver as a coating material are being reported by many authors in their literature [14,15,36].

In this paper, a new simple reconfigurable surface-plasmon-based filter/sensor using D-shaped photonic crystal fiber is proposed. The proposed structure can be used as a polarization filter without a filling analyte and, at the same time, it can act as a sensor when it is filled with an analyte, which is interested in being sensed. The asymmetrical core design helps enhance the birefringence and, hence, it exhibits the difference in X and Y-polarization propagation of a fundamental mode. This principle helps the fiber realize it as a polarization filter. By using the phase-matching condition between the surface plasmon mode and either one polarization mode of the fiber, particular polarization can be filtered. A tiny layer of silver is placed on the flat surface of the D-shaped fiber with a small half-circular opening to activate the plasmon modes. Various structural characteristics, such as silver thickness, air hole size and shape, and pitch size, are tuned to improve the filter's filtering efficiency. As the second part of our proposed work, the same structure is extended as a sensor by filling analytes in the outer circle of the proposed filter design. The performances of both polarization filter and RI sensor are analyzed using the finite element method (FEM). The rest of the paper is organized as follows. Section 2 presents the design of the D-shaped PCF. Section 3 discusses the results for the proposed PCF as (Section 3.1) a filter and (Section 3.2) a sensor, and finally, the conclusion is bestowed in Section 4.

2. Design of D-Shaped Photonic Crystal Fiber

Figure 1 shows the schematic diagram of a D-shaped PCF with a silver layer coated on its top surface. In the proposed structure, the air holes are arranged in a hexagonal lattice with 3.5 μm lattice size. The air holes that have different diameters ranging from 1.5 to 3.3 μm form a single-core PCF. Here, a unique PCF structure with minimal number of air holes is used for higher light confinement. Figure 1 illustrates the schematic diagram of the proposed reconfigurable D-shaped PCF filter/sensor. D-shaped PCF is formed by slicing the PCF at 8 μm from the core center. The 15 nm silver layer is coated on the top surface of the PCF, which acts as the noble metal. In the proposed structure, the air holes are arranged in a hexagonal lattice with a 3.5 μm lattice size forming a single-core PCF. All air holes have different diameters, ranging from d1 to d6. The values of these diameters are as follows: d1 = 2.3 μm, d2= 3.3 μm, d3 = 1.5 μm, d4 = 1.84 μm, d5 = 2 μm and d6 = 1.7 μm. The pitch value (Λ_1) between the d2, d3 and d6 air holes is 7 μm. The pitch values Λ_2 = 9 μm, Λ_3 = 15 μm and Λ_4 = 12 μm are maintained between d4, d5 and top d4, respectively. The air holes d1, d2 and d3 are located at 3.5 μm from the center in the Y axis. Air holes d4 are located at 4.5 μm from the center in the Y axis. Air holes d6 and d3 are located in the Y axis from the center at 7 μm and 6.5 μm, respectively. The chosen diameter of the PCF is 17.5 μm. A minimal number of air holes were preferred for higher confinement and also to obtain a unique structure. Here, the missing air holes are used to create an asymmetric core that helps achieve single-mode polarization filtering by increasing the birefringence. A micro-opening is created in the middle of the top side with a diameter of 3.4 μm to influence the phase-matching condition between the core mode and the plasmon mode using analyte. Furthermore, this micro-opening also provides two channels, allowing for improved interaction between the core and plasmon modes.

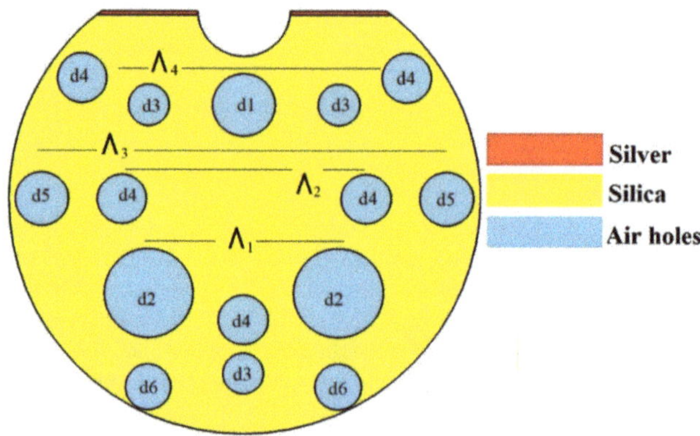

Figure 1. Proposed silver lining D-shaped PCF (for both filter and sensor applications).

Because of the short resonance peak when compared to gold and its resonance capability from visible to infrared wavelength range, silver is preferred as the plasmonic material [36,37]. A thin layer of silver with thickness of 15 nm is deposited on these two pathways. SPR takes place through this silver layer. Light propagating through the core induces plasmons to develop in the silver layer. The maximum energy couples from the core mode to the plasmon mode at the phase-matching condition. This coupling creates a decline in energy in the core mode, which attenuates one of the polarized lights while allowing the other due to its birefringence behavior. As a result, the PCF can act as a polarization filter. The scattering boundary condition is established to minimize scattering losses. The performance evaluation of the proposed system is carried out using the finite element method (FEM). The proposed filter structure can be fabricated using the stack-and-draw approach [38] and mechanical side polishing methods [39]. The chemical vapor deposition method is used to deposit the silver lining [40].

Pure silica is used as a core material in this PCF-based Polarization filter, and its refractive index value is calculated using the Sellmeier equation [41]:

$$n(\lambda)^2 = \frac{A_1\lambda^2}{\lambda^2 - \lambda_1^2} + \frac{A_2\lambda^2}{\lambda^2 - \lambda_2^2} + \frac{A_3\lambda^2}{\lambda^2 - \lambda_3^2} \tag{1}$$

where, $A_1 = 0.6961663$, $A_2 = 0.4079426$, $A_3 = 0.897479$, $\lambda_1 = 0.068404$, $\lambda_2 = 0.1162414$, $\lambda_3 = 9.896161$ are the Sellmeier coefficients.

Dielectric constant value of silver is calculated using the Drude dispersion model [42,43]:

$$\varepsilon_{Ag} = 1 - \frac{\lambda^2\lambda_C}{\lambda_p^2(\lambda_c + i\lambda)} \tag{2}$$

where collision wavelength $\lambda_c = 17.614$ μm, Plasma wavelength $\lambda_p = 0.14541$ μm, and λ being the free space wavelength.

Confinement loss(α) of the filter can be determined numerically using the following equation [44]:

$$\alpha(\text{dB/cm}) = 8 \cdot 686 \times \frac{2\pi}{\lambda} \times \text{Im}(n_{eff}) \times 10^4 \tag{3}$$

where $\text{Im}(n_{eff})$—imaginary part of effective mode index.

Figure 2 shows the birefringence value of the proposed fiber without silver lining and analyte. The observed birefringence values are 0.73×10^{-4}, 1.08×10^{-4} and 1.76×10^{-4} at wavelengths 1.55 μm, 1.8 μm and 2.2 μm, respectively. These birefringence values are of the order of 2 to 3 times more than the normal PCF. This ensures the SPR at different wavelengths for the X and Y-polarization mode. Consequently, this enables the polarization filter function and sensing behavior at a particular polarization using the proposed PCF.

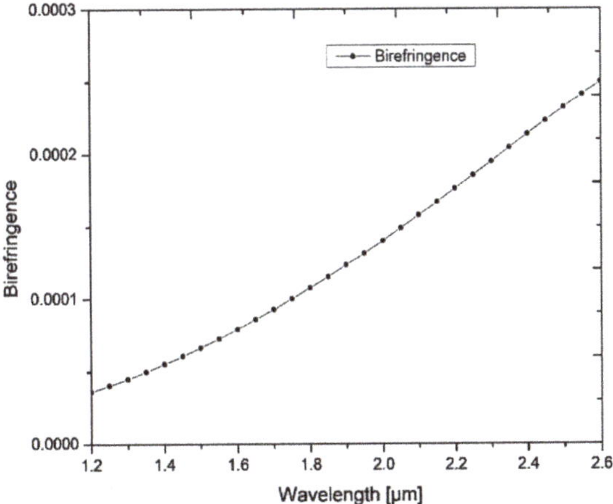

Figure 2. The birefringence vs wavelength of the proposed fiber without silver lining and analyte.

3. Results and Discussion

3.1. SPR Based Filter

3.1.1. Optimization of the Structural Parameters

To obtain an optimal filter characteristic, the effects of structural parameters such as size and shape of the air holes, lattice size, and silver layer thickness are examined. By changing one of these parameters iteratively while keeping the others constant, the performance of the proposed PCF filter design is examined.

3.1.2. Varying Size of Air Holes

The filter characteristics of the proposed PCF structure measured for different air holes sizes are displayed in Figure 3. The confinement loss is realized using the surface resonance and it is measured for different air holes diameter values as illustrated in Figure 4a,b for both X-polarization and Y-polarization. As the air holes size decreases (high lattice), the sharp confinement loss peak shifts towards shorter wavelengths. The effective area of the core is increased as the air holes sizes are reduced. At phase-matching conditions, more energy couples from core mode to plasmon mode and it creates more confinement loss at core mode. A confinement loss of 713 dB/cm has been reported at 1.98 µm for X-polarized mode with a pitch of 3.5 µm. For Y-polarization the observed confinement loss is 426 dB/cm at 1.92 µm.

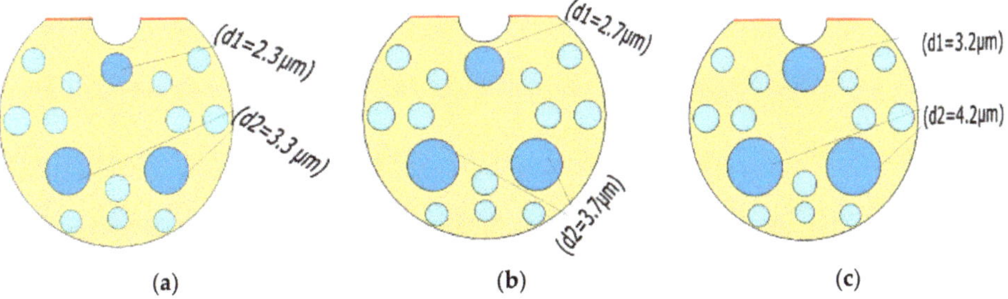

Figure 3. D-shaped PCF based polarization filter with different air hole's sizes (**a**) air holes dia d1 = 2.3 µm and d2 = 3.3 µm (high lattice) (**b**) air holes dia d1 = 2.7 µm and d2 = 3.7 µm (Mid lattice) (**c**) air holes dia d1 = 3.2 µm and d2 = 4.2 µm (low lattice).

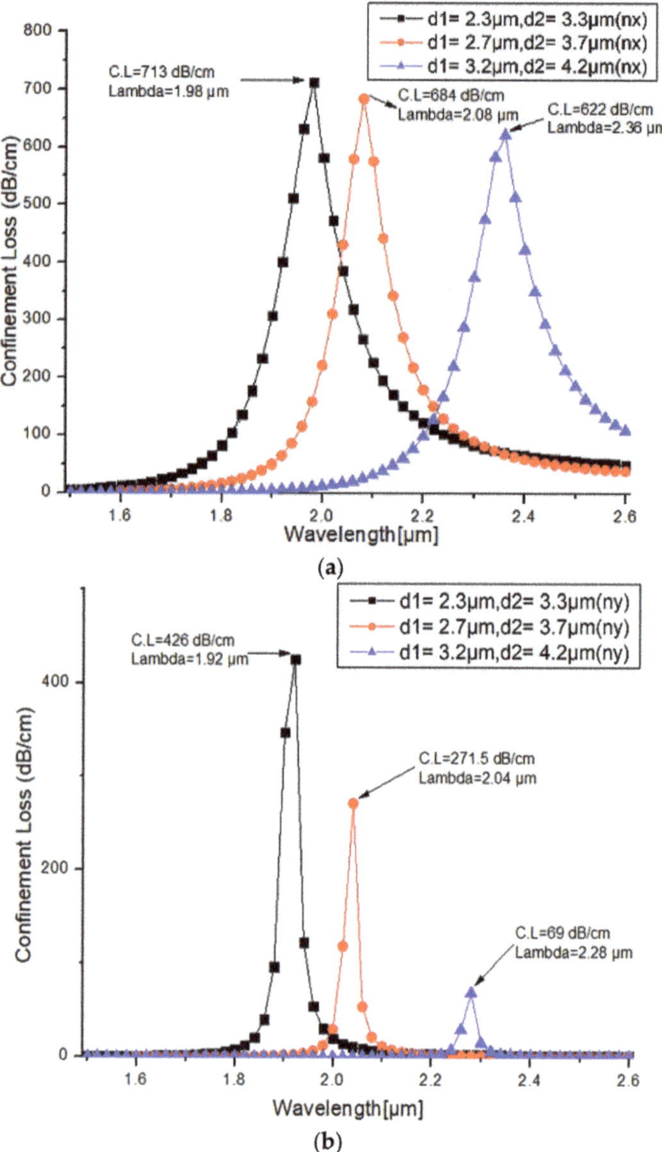

Figure 4. Confinement loss for the different pitch values (**a**) X-polarized mode (**b**) Y-polarized mode.

3.1.3. Varying Shape of the Air Holes

To find the optimal filter structure, different airhole shapes were chosen, from circular to ones depicted in Figure 5a–c. As the diameter of the air holes increases, the effective area of the core mode decreases, causing the resonance peak to shift toward shorter wavelengths. Figure 5a shows PCF with standard circular air holes. Additionally, the modified three elliptical air holes and four elliptical air holes are shown in Figure 5b,c respectively.

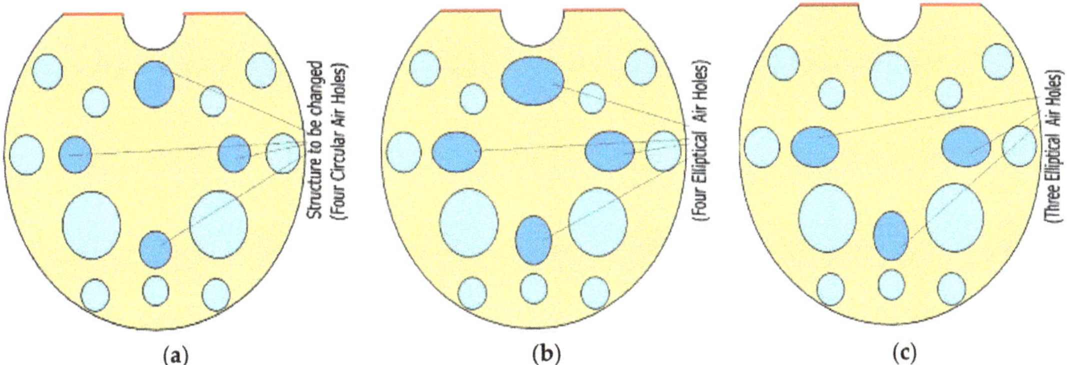

Figure 5. D-shaped PCF-based polarization filter with different air hole structure (**a**) all-circular air holes (**b**) with three elliptical air holes (**c**) with four elliptical air holes.

Figure 6a,b show the comparison of confinement loss peaks for the X-polarization and Y-polarization with different structures of air holes. As shown in Figure 5, circular air holes, three elliptical air holes and four elliptical air holes are preferred to improve the structure's efficiency and thereby find the optimal structure. For wavelengths ranging from 1.5 μm to 2.6 μm, numerical analysis was carried out with a step value of 0.02. From this analysis, the observed highest confinement loss for circular air holes is about 713 dB/cm at an operating wavelength of 1.98 μm. For three elliptical holes, the observed confinement loss is 584 dB/cm at the operating wavelength of 2.38 μm. For four elliptical holes, the confinement loss is 487 dB/cm at an operating wavelength of 2.4 μm. From this graph of X-polarization, the circular air holes have the highest confinement loss compared to three elliptical holes and four elliptical holes; the difference is fairly high. The circular air holes are only carried for further analysis due to their superior performance in terms of their confinement loss.

3.1.4. Varying the Thickness of Silver

The silver lining thickness has an impact on the efficiency of the plasmonic based PCF filter by changing the plasmonic resonance. As per Chao Liu, et.al. [45], 20 nm thickness of silver provides a better sensitivity among the thickness range from 20 to 50 nm. It clearly states that less thickness provides a better surface plasmon resonance compared to the higher thickness level and hence it is decided to carry out further research to improve the sensitivity of the sensor by changing the silver coating thickness in the range between 5 nm and 15 nm with a step size of 5 nm. Out of this research, it is found and reported here that 15 nm thickness of Ag provides good confinement loss and wavelength sensitivity as shown in Figure 7a,b. Under phase matching conditions, the observed maximum loss is 713 dB/cm for the X-polarized mode at the operating wavelength of 1.98 μm. Also, the maximum loss of 426 dB/cm is recorded at the operating wavelength of 1.92 μm for Y-polarized mode.

3.1.5. Transmission Characteristics and Dispersion Relation

For the optimized structure of the filter, the light coupling characteristics are investigated over a wavelength range from 0.8 to 2.2 μm in both X and Y-polarizations. The maximum energy of the core mode is transferred to SPP mode at the phase-matching condition, and the energy of the core mode rapidly decreases. As a result, it will attenuate one of the polarized lights while allowing the other to go through. Figure 8a,b illustrate the electric field distributions of X-polarized mode at wavelength 1.98 μm and Y-polarized mode at wavelength 1.92 μm.

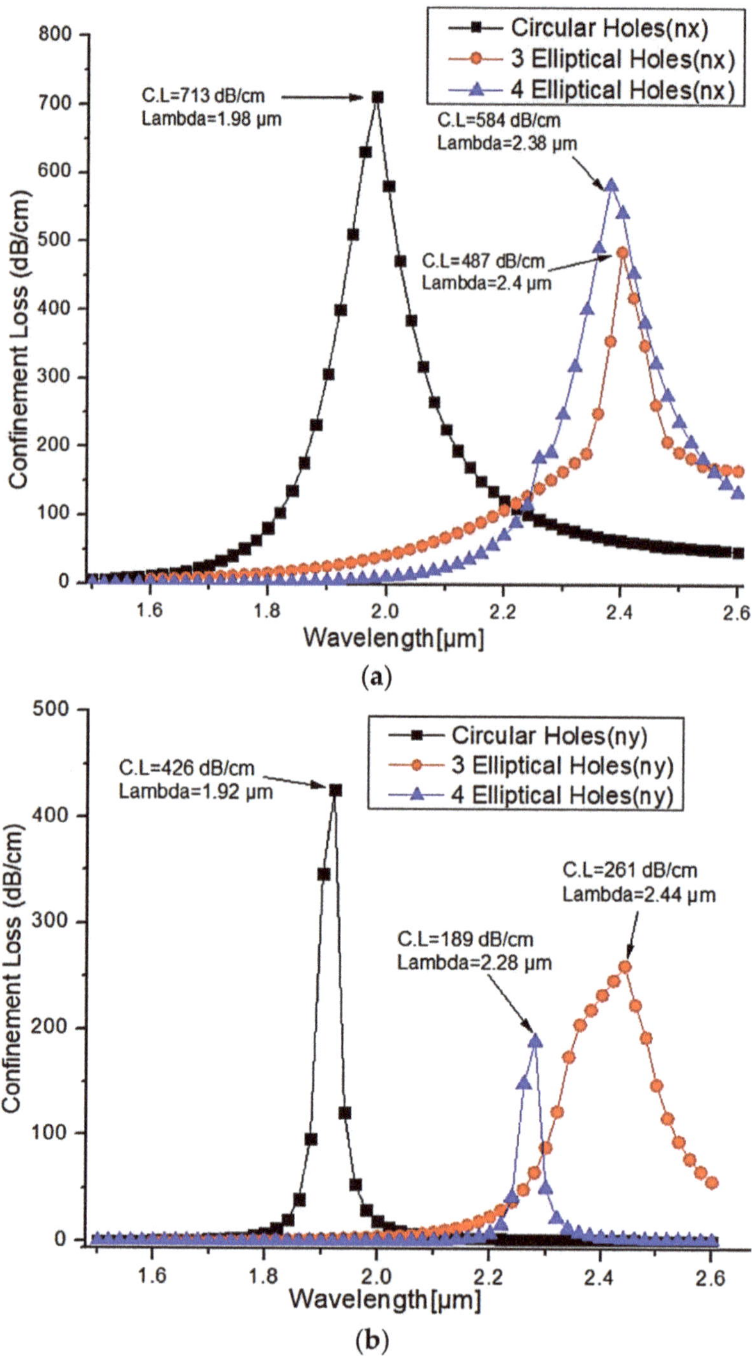

Figure 6. The confinement loss for the different shapes of air holes in (**a**) X-polarized mode (**b**) Y-polarized mode.

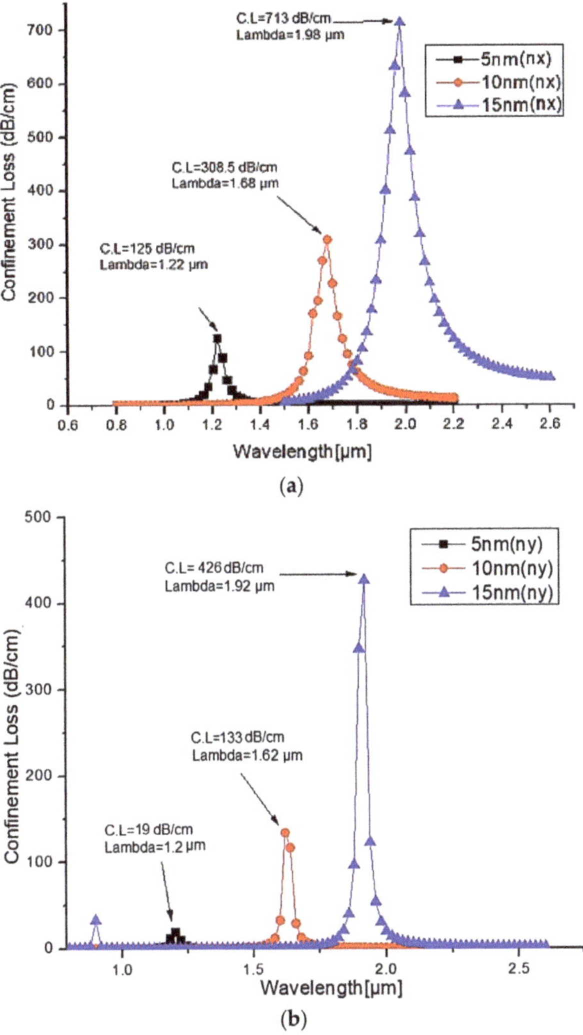

Figure 7. Characteristics of confinement loss for various silver layer thicknesses (5, 10 and 15 nm) (**a**) X-polarized mode (**b**) Y-polarized mode.

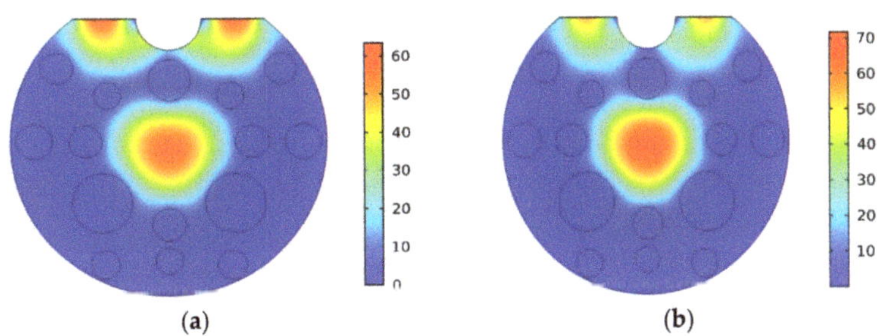

Figure 8. Modal field distribution (**a**) Resonance condition for X-polarization at 1.98 µm (**b**) Resonance condition for Y-polarization at 1.92 µm.

The dispersion relation between the real and imaginary values of the effective mode index (n_{eff}) is depicted in Figure 9. As shown in Figure 9, the real part of the core-guided mode decreases when the wavelength increases. The fiber dispersion characteristic is defined by the real values of n_{eff} [46], and the confinement loss is calculated through imaginary values of n_{eff} [47]. As depicted in Figure 9, the X-polarized mode couples with the plasmon mode, and hence, the core mode imaginary RI value increases at 1.98 μm. The observed confinement loss is 713 dB/cm at this wavelength, whereas the Y-polarized mode couples with the plasmon mode and has a confinement loss of 426 dB/cm at the wavelength of 1.92 μm. As per the result, at the wavelength of 1.98 μm, only the Y-polarized mode propagates, and the X-polarized mode is attenuated, and at the wavelength of 1.92 μm, only the X-polarized mode propagates, and the Y-polarized mode is attenuated. As a result, the proposed structure functions as a polarization filter by removing one of the polarized light waves at a particular wavelength.

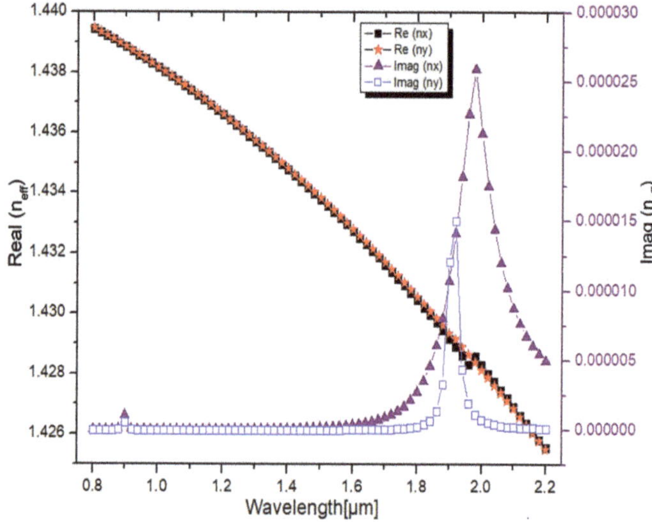

Figure 9. Real and imaginary parts of the PCF filter core guided mode as a function of wavelength.

3.2. SPR Based Sensor

The optimized SPR-based polarization filter structure extends to sensing application by filling an analyte in the outer layer of the sensor. When the refractive index of the analyte changes, the structure resonant wavelength will shift. For the sensing purpose, the resonance wavelength shift (wavelength sensitivity) scheme is preferred because it results in a high level of sensitivity [48]. A schematic representation of the plasmonic sensor is shown in Figure 10. A PML layer is used to reduce scattering losses. The analyte's indices influence the amount of light coupled into the silver layer (surface plasmon mode) and thus deciding light propagation through the core.

Figure 11 depicts the real and imaginary values of the core-guided mode for the analyte RI of 1.36. As shown in Figure 11, the real part of the core-guided mode decreases as the wavelength increases. The fiber's core-guided mode is coupled with the plasmon mode at a wavelength of 2.46 μm. As a result, maximum energy is coupled from the core mode to the plasmon mode, causing fundamental mode loss. At the resonance condition, the imaginary value of the core-guided mode reaches its maximum value. The performance characteristic of a sensor is defined by its sensitivity, and thus, sensitivity needs to be analyzed. Sensitivity (wavelength shift) of a sensor is calculated [49] as follows,

$$S_\lambda = \frac{\Delta\lambda_{Peak}}{\Delta n_a}(nm/RIU) \quad (4)$$

$\Delta\lambda_{peak}$–resonance wavelength shift, Δn_a—analyte RI value shift.

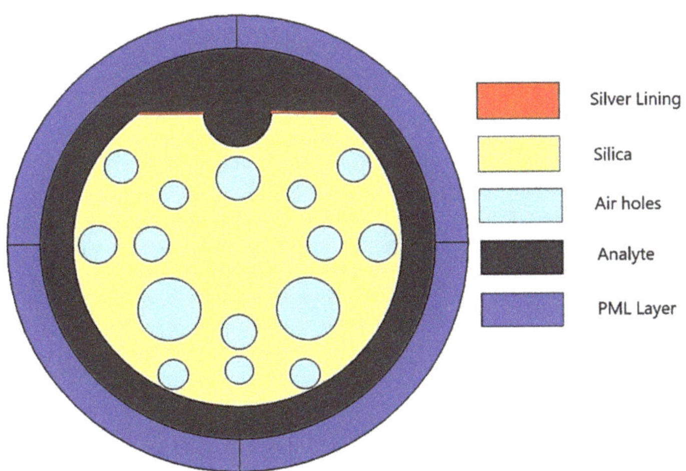

Figure 10. Graphical representation of the SPR based sensor structure.

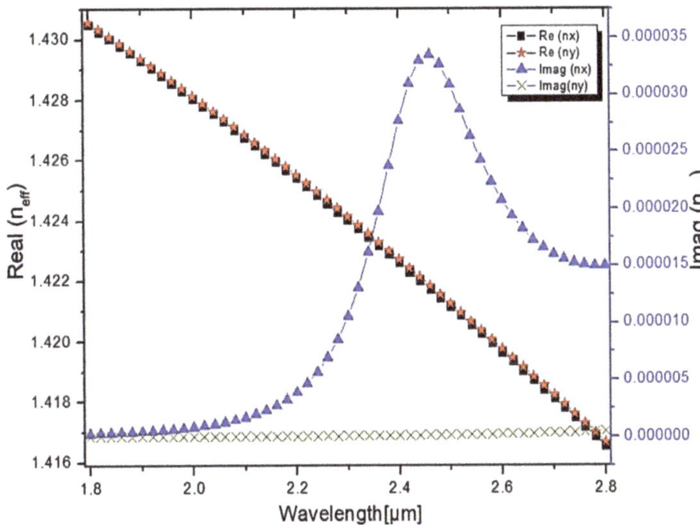

Figure 11. Real and Imaginary values of 1.36 RI.

For the proposed sensor, a numerical investigation was carried out over a wavelength ranging from 1.8 μm to 2.8 μm, with a step value of about 0.02 μm. Additionally, for the analytes, the refractive indices ranging from 1.34 to 1.37 in the step of 0.01 were taken for the analysis. Figures 12 and 13 illustrate the confinement loss measured in dB/cm for both the X and Y-polarization. As shown in Figure 12, the observed confinement losses at the X-polarization are 505 dB/cm, 602 dB/cm, 744 dB/cm, 1048 dB/cm at wavelengths of

2.38 μm, 2.42 μm, 2.46 μm, 2.52 μm, for the analyte refractive indices of 1.34, 1.35, 1.36, 1.37, respectively. From this analysis, the maximum confinement loss of about 1048 dB/cm is observed for the analyte refractive index of 1.37 at the operating wavelength of 2.52 μm, whereas in the Y-polarization, the observed confinement losses are very low, as depicted in Figure 12, and for the refractive indices of 1.34, 1.35, 1.36, 1.37, the recorded maximum values at the wavelength of 2.8 μm are 9.2 dB/cm, 9.5 dB/cm, 10 dB/cm, 32 dB/cm, respectively. Figure 13 is an extracted image from Figure 12 to display the confinement loss at the Y-polarization. From this analysis, it is clearly shown that the proposed PCF structure can be used as a refractive index sensor by shifting the confinement loss with respect to the wavelength in the X-polarization for the different analyte indices.

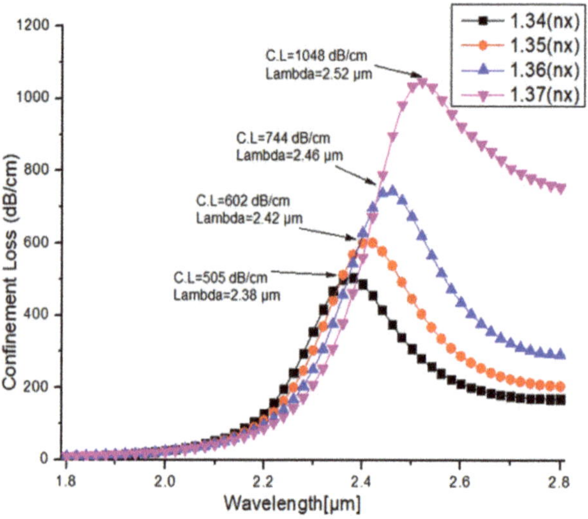

Figure 12. Confinement losses in both X and Y-polarizations for the different RI ranging from 1.34 to 1.37 with step value of 0.01.

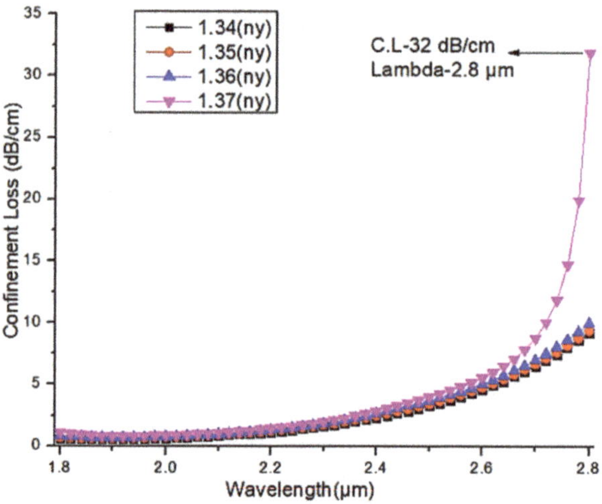

Figure 13. Confinement loss in Y-polarization with respect to wavelength (extracted from Figure 12 for more clarity).

The sensitivity of the D-shaped silver lining plasmonic sensor is calculated using Equation (4). The sensitivity of the proposed system over the RI ranging from 1.34 to 1.37 is depicted in Figure 14. The observed sensitivity from RI 1.34 to 1.36 is 4000 nm/RIU, and the sensitivity is improved to 5000 nm/RIU for the RI ranging from 1.36 to 1.37. This sensor can be implemented to detect the small changes in the bio analytes and industry chemicals. The comparison of the proposed work with contemporary literature works is as depicted in Table 1. It shows a good wavelength sensitivity in the proposed work over the RI range from 1.34 to 1.37. This RI range covers the sensing of many industrial solvents and bio-analytes [50] RIs.

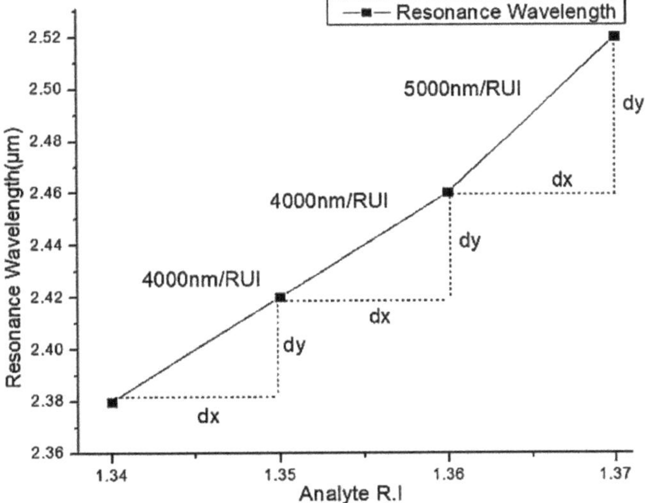

Figure 14. Sensor sensitivity measurement using resonance wavelength shift with reference to analyte RI changes.

Table 1. Comparison of our proposed work with contemporary literatures.

Year/Ref	Plasmonic Material Used	Application Demonstrated	Sensitivity [nm/RIU]	RI Range
2015/[37]	Silver/Graphene	Sensor	3000	1.46–1.49
2017/[51]	Gold	Sensor	4000	1.33–1.37
2018/[10]	Gold	Sensor	4600	1.33–1.38
2018/[52]	Gold	sensor	9000	1.34–1.37
2019/[15]	Silver/Graphene	Sensor	3750	1.33–1.35
2019/[53]	Silver/Graphene	Sensor	833.33	1.30–1.34
2020/[13]	Silver/Gold	Sensor	3083	1.33–1.366
2020/[54]	Gold/Titanium dioxide	Sensor	4782	1.49–1.54
2021/[55]	Gold/polydimethylsiloxane	Sensor	1371	1.33–1.34
2022/[56]	Silver	Sensor	4100	1.38–1.41
2022/[57]	Silver	Sensor	1932	1.25–1.30
2022/[58]	Gold	Sensor	3000	1.40–1.46
2022/[59]	Gold	Sensor	3300	1.35–1.40
Proposed work	Silver	Filter/sensor	5000	1.34–1.37

Compared to the literature works [10,13,15,51,55,59], the proposed work provides superior wavelength sensitivity of 5000 nm/RIU, approximately, for the RI ranges from 1.33 to 1.40. The reported work in Ref. [52] covered exactly the RI range of 1.34–1.37, similar to this proposed work, with a sensitivity of about 9000 nm/RIU. However, the PCF reported in Refs. [10,52] had the air hole at the axis of the fiber, with a dual channel on either side of the fiber axis to sense the change in RI values. Compared to these literature works, the proposed work is very simple in design and less complex to couple the light for the polarization-sensitive sensors. Contrarily, the literature works [10,52] will create losses while coupling the light from an external single-mode fiber due to the presence of the airhole at the center. Moreover, this air hole at the fiber axis will create complexity to couple the light with proper polarization in polarization-sensitive sensors.

Figure 15 depicts the experimental arrangement for the proposed reconfigurable D-shaped SPR filter/sensor. A light source with a wide spectrum is launched into the proposed fiber through SMF for better coupling. Depending on the analyte being filled or unfilled around the proposed fiber, this setup can act as a polarization filter or sensor. As a sensor, the addition of a polarizer gives better visibility regarding the confinement loss under a particular polarization. As per the RI of the analyte, the fiber mode loss varies at a particular wavelength, and it can be recorded on an optical spectrum analyzer (OSA) and computer connected to it. By analyzing the data from the computer, the unknown analyte RI can be identified through the wavelength shift of peak loss in SPR between the core mode and the plasmon mode.

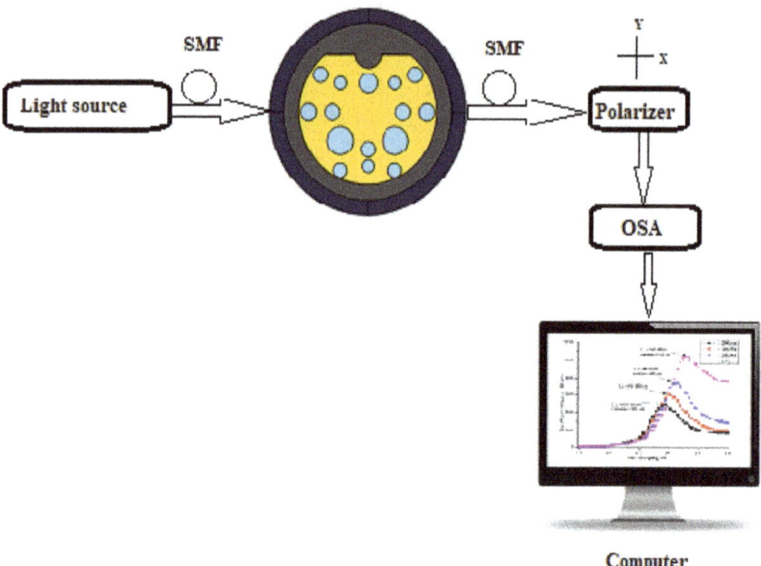

Figure 15. Experimental arrangement for the proposed reconfigurable D-shaped SPR filter/sensor.

4. Conclusions

SPR based silver lining reconfigurable D-shaped photonic crystal fiber is proposed to filter out the single-polarization and to sense the RI of analytes. A tiny layer of silver is deposited over the flat surface of the D-shaped fiber with a small half-circular opening to activate the plasmon modes. Structural optimization has been done through FEM to obtain a single polarization filter and the performance parameters are measured. For the application of PCF as a filter without adding analyte, the maximum confinement loss of 713 dB/cm is observed in X-polarization at the wavelength of 1.98 µm and for Y-polarization the maximum confinement loss is observed as 426 dB/cm at 1.92 µm. This property allows the proposed structure to act as X or Y-polarization filter at the wavelength of 1.98 μm and 1.92 μm respectively. This PCF filter can be used as a sensor by adding an analyte outside this filter structure. With a maximum sensitivity of 5000 nm/RIU at X-polarization, this sensor can detect small refractive index (RI) variation of analytes ranging from 1.34 to 1.37. This sensor can be used for industrial and medical applications to measure chemical concentrations and disease levels using sample analytes. In future, research can be focused towards finding a good improvement in the dynamic range of sensing properties by extending the same structure with different materials like chalcogenide, Fluoride glass and Bismuth oxide instead of silica.

Author Contributions: Conceptualization, methodology, investigation and original draft preparation, S.S.; software, and data curation, J.D.; validation A.S.R.; formal analysis A.S.; Visualization, review and editing, S.I. All authors have read and agreed to the published version of the manuscript.

Funding: This research received no external funding.

Data Availability Statement: Not applicable.

Acknowledgments: The authors thankfully acknowledge the financial support rendered by the leadership and management of Vellore Institute of Technology Chennai under its research promotion scheme.

Conflicts of Interest: The authors declare no conflict of interest.

References

1. Hasan, M.; Abdulrazak, L.F.; Paul, B.K.; Al-Zahrani, F.A.; Ahmed, K. Novel shaped solid-core photonic crystal fiber for the numerical study of nonlinear optical properties. *Opt. Quantum Electron.* **2022**, *54*, 139. [CrossRef]
2. Rahaman, E.; Hossain, M.; Mondal, H.S.; Saha, R.; Muntaseer, A.S. Theoretical analysis of large negative dispersion photonic crystal fiber with small confinement loss. *Appl. Opt.* **2020**, *59*, 8925–8931. [CrossRef]
3. Imam, F.; Biswas, S.; Shahjahan; Khatun, H.; Ahmmed, R.; Bulbul, A.A.-M. Design and numerical analysis of a dispersion-flattened fabrication-friendly PCF-based THz waveguide. *J. Opt.* **2021**, *51*, 371–378. [CrossRef]
4. Rahaman, E.; Mondal, H.S.; Hossain, B.; Hossain, M.; Ahsan, S.; Saha, R. Simulation of a highly birefringent photonic crystal fiber in terahertz frequency region. *SN Appl. Sci.* **2020**, *2*, 1435. [CrossRef]
5. Kuiri, B.; Dutta, B.; Sarkar, N.; Santra, S.; Mandal, P.; Mallick, K.; Patra, A.S. Ultra-low loss polymer-based photonic crystal fiber supporting 242 OAM modes with high bending tolerance for multimode THz communication. *Results Phys.* **2022**, *36*, 105465. [CrossRef]
6. Yang, J.; Zhou, L.; Che, X.; Huang, J.; Li, X.; Chen, W. Photonic Crystal Fiber Methane Sensor Based on Modal Interference with an Ultraviolet Curable Fluoro-Siloxane Nano-Film Incorporating Cryptophane A. *Sens. Actuators B Chem.* **2016**, *235*, 717–722. [CrossRef]
7. Cheng, T.; Duan, Z.; Liao, M.; Gao, W.; Deng, D.; Suzuki, T.; Ohishi, Y. A Simple All-Solid Tellurite Microstructured Optical Fiber. *Opt. Express* **2013**, *21*, 3318–3323. [CrossRef]
8. Wong, W.C.; Chan, C.C.; Chen, L.H.; Li, T.; Lee, K.X.; Leong, K.C. Polyvinyl Alcohol Coated Photonic Crystal Optical Fiber Sensor for Humidity Measurement. *Sens. Actuators B Chem.* **2012**, *174*, 563–569. [CrossRef]
9. Shafkat, A. Analysis of a gold coated plasmonic sensor based on a duplex core photonic crystal fiber. *Sens. Bio-Sens. Res.* **2020**, *28*, 100324. [CrossRef]
10. Hasan, M.R.; Akter, S.; Rifat, A.A.; Rana, S.; Ahmed, K.; Ahmed, R.; Subbaraman, H.; Abbott, D. Spiral photonic crystal fiber-based dual-polarized surface plasmon resonance biosensor. *IEEE Sens. J.* **2017**, *18*, 133–140. [CrossRef]
11. Yang, X.; Lu, Y.; Liu, B.; Yao, J. Simultaneous measurement of refractive index and temperature based on SPR in D-shaped MOF. *Appl. Opt.* **2017**, *56*, 4369–4374. [CrossRef]
12. Rahman, M.; Molla, A.; Paul, A.K.; Based, A.; Rana, M.; Anower, M. Numerical investigation of a highly sensitive plasmonic refractive index sensor utilizing hexagonal lattice of photonic crystal fiber. *Results Phys.* **2020**, *18*, 103313. [CrossRef]
13. Yasli, A.; Ademgil, H.; Haxha, S.; Aggoun, A. Multi-Channel Photonic Crystal Fiber Based Surface Plasmon Resonance Sensor for Multi-Analyte Sensing. *IEEE Photonics J.* **2020**, *12*, 6800515. [CrossRef]
14. Chu, S.; Nakkeeran, K.; Abobaker, A.M.; Aphale, S.S.; Sivabalan, S.; Babu, P.R.; Senthilnathan, K. A Surface Plasmon Resonance Bio-Sensor Based on Dual Core D-Shaped Photonic Crystal Fiber Embedded with Silver Nanowires for Multisensing. *IEEE Sens. J.* **2021**, *21*, 76–84. [CrossRef]
15. Yasli, A.; Ademgil, H. Effect of plasmonic materials on photonic crystal fiber-based surface plasmon resonance sensors. *Mod. Phys. Lett. B* **2019**, *33*, 1950157. [CrossRef]
16. Azman, M.F.; Mahdiraji, G.A.; Wong, W.R.; Aoni, R.A.; Adikan, F.R.M. Design and fabrication of copper-filled photonic crystal fiber-based polarization filters. *Appl. Opt.* **2019**, *58*, 2068–2075. [CrossRef]

17. Jabin, A.; Ahmed, K.; Rana, J.; Paul, B.K.; Islam, M.; Vigneswaran, D.; Uddin, M.S. Surface Plasmon Resonance Based Titanium Coated Biosensor for Cancer Cell Detection. *IEEE Photonics J.* **2019**, *11*, 3700110. [CrossRef]
18. Heikal, A.; Hameed, M.; Obayya, S.S.A. Improved trenched channel plasmonic waveguide. *J. Lightwave Technol.* **2013**, *31*, 2184–2191. [CrossRef]
19. Burstein, E.; Chen, W.P.; Chen, Y.J.; Hartstein, A. Surface polaritons—propagating electromagnetic modes at interfaces. *J. Vac. Sci. Technol.* **1974**, *11*, 1004. [CrossRef]
20. Selvendran, S.; Sivanantharaja, A.; Yogalakshmi, S. A highly sensitive Bezier polygonal hollow core photonic crystal fiber biosensor based on surface plasmon resonance. *Optik* **2018**, *171*, 109–113. [CrossRef]
21. Yogalakshmi, S.; Selvendran, S.; Raja, A.S. Design and analysis of a photonic crystal fiber-based polarization filter using surface plasmon resonance. *Laser Phys.* **2016**, *26*, 056201. [CrossRef]
22. Xu, Q.; Li, K.; Copner, N.; Lin, S. An Ultrashort Wavelength Multi/Demultiplexer via Rectangular Liquid-Infiltrated Dual-Core Polymer Optical Fiber. *Materials* **2019**, *12*, 1709. [CrossRef] [PubMed]
23. Rajeswari, D.; Raja, A.S.; Selvendran, S. Design and analysis of polarization splitter based on dual-core photonic crystal fiber. *Optik* **2017**, *144*, 15–21. [CrossRef]
24. Yang, X.; Lu, Y.; Liu, B.; Yao, J. Tunable polarization filter based on high-birefringence photonic crystal fiber filled with silver wires. *Opt. Eng.* **2017**, *56*, 077108. [CrossRef]
25. Sharma, M.; Mishra, S.K.; Ung, B. Ultra-sensitive and large dynamic range refractive index sensor utilizing annular core photonic crystal fiber. *Spie* **2019**, *11123*, 1112305.
26. Zhou, X.; Li, S.; Cheng, T.; An, G. Design of offset core photonic crystal fiber filter based on surface plasmon resonance. *Opt. Quantum Electron.* **2018**, *50*, 157. [CrossRef]
27. Yang, X.; Lu, Y.; Liu, B.; Yao, J. Polarization characteristics of high-birefringence photonic crystal fiber selectively coated with silver layers. *Plasmonics* **2018**, *13*, 1035–1042. [CrossRef]
28. Liu, C.; Wang, L.; Wang, F.; Xu, C.; Liu, Q.; Liu, W.; Yang, L.; Li, X.; Sun, T.; Chu, P.K. Tunable single-polarization bimetal-coated and liquid-filled photonic crystal fiber filter based on surface plasmon resonance. *Appl. Opt.* **2019**, *58*, 6308–6314. [CrossRef]
29. Yan, X.; Guo, Z.; Cheng, T.; Li, S. A novel gold-coated PCF polarization filter based on surface plasmon resonance. *Opt. Laser Technol.* **2020**, *126*, 106125. [CrossRef]
30. Lavanya, A.; Geetha, G. Broadband polarization filter based on hybrid silver-graphene coated pentagonal photonic crystal fiber. *Opt. Quantum Electron.* **2021**, *53*, 117. [CrossRef]
31. Guo, Y.; Li, J.; Li, S.; Zhang, S.; Liu, Y. Broadband single-polarization filter of D-shaped photonic crystal fiber with a micro-opening based on surface plasmon resonance. *Appl. Opt.* **2018**, *57*, 8016–8022. [CrossRef]
32. Almewafy, B.H.; Areed, N.F.F.; Hameed, M.F.O.; Obayya, S.S.A. Modified D-shaped SPR PCF polarization filter at telecommunication wavelengths. *Opt. Quantum Electron.* **2019**, *51*, 193. [CrossRef]
33. Shima, R.A.; Mollah, A.; Ali, Y. Au-ITO deposited D-shaped photonic crystal fiber polarizer with a micro-opening based on surface plasmon resonance. *Optik* **2020**, *224*, 165489. [CrossRef]
34. Johnson, P.B.; Christy, R.W. Optical constants of the noble metals. *Phys. Rev.* **1972**, *6*, 4370. [CrossRef]
35. Rifat, A.A.; Ahmed, R.; Mahdiraji, G.A.; Adikan, F.R.M. Highly Sensitive D-Shaped Photonic Crystal Fiber-Based Plasmonic Biosensor in Visible to Near-IR. *IEEE Sens. J.* **2017**, *17*, 2776–2783. [CrossRef]
36. Hossain, B.; Islam, S.R.; Hossain, K.T.; Abdulrazak, L.F.; Sakib, N.; Amiri, I. High sensitivity hollow core circular shaped PCF surface plasmonic biosensor employing silver coat: A numerical design and analysis with external sensing approach. *Results Phys.* **2020**, *16*, 102909. [CrossRef]
37. Rifat, A.A.; Mahdiraji, G.A.; Chow, D.M.; Shee, Y.G.; Ahmed, R.; Adikan, F.R.M. Photonic Crystal Fiber-Based Surface Plasmon Resonance Sensor with Selective Analyte Channels and Graphene-Silver Deposited Core. *Sensors* **2015**, *15*, 11499–11510. [CrossRef]
38. Zhang, N.; Li, K.; Cui, Y.; Wu, Z.; Shum, P.P.; Auguste, J.-L.; Dinh, X.Q.; Humbert, G.; Wei, L. Ultra-sensitive chemical and biological analysis via specialty fibers with built-in micro structured optofluidic channels. *Lab Chip* **2018**, *18*, 655–661. [CrossRef]
39. Huang, T. Highly sensitive SPR sensor based on D-shaped photonic crystal fiber coated with indium tin oxide at near-infrared wavelength. *Plasmonics* **2017**, *12*, 583–588. [CrossRef]
40. Boehm, J.; François, A.; Ebendorff-Heidepriem, H.; Monro, T. Chemical deposition of silver for the fabrication of surface plasmon micro structured optical fibre sensors. *Plasmonics* **2011**, *6*, 133–136. [CrossRef]
41. Sellmeier, W. To explain the abnormal color sequence in the spectrum of some substances. *Ann. Phys.* **1870**, *219*, 272–282. [CrossRef]
42. Jiao, S.; Gu, S.; Fang, H.; Yang, H. Analysis of Dual-Core Photonic Crystal Fiber Based on Surface Plasmon Resonance Sensor with Segmented Silver Film. *Plasmonics* **2019**, *14*, 685–693. [CrossRef]
43. Chen, W.; Thoreson, M.D.; Ishii, S.; Kildishev, A.V.; Shalaev, V.M. Ultra-thin ultra-smooth and low-loss silver films on a germanium wetting layer. *Opt. Express* **2010**, *18*, 5124–5134. [CrossRef]
44. Rifat, A.A.; Mahdiraji, G.A.; Sua, Y.M.; Shee, Y.G.; Ahmed, R.; Chow, D.; Adikan, F.R.M. Surface plasmon resonance photonic crystal fiber biosensor: A practical sensing approach. *IEEE Photonics Technol. Lett.* **2015**, *27*, 1628–1631. [CrossRef]
45. Liu, C.; Wang, J.; Jin, X.; Wang, F.; Yang, L.; Lv, J.; Fu, G.; Li, X.; Liu, Q.; Sun, T.; et al. Near-infrared surface plasmon resonance sensor based on photonic crystal fiber with big open rings. *Optik* **2020**, *207*, 164466. [CrossRef]

46. Paul, B.K.; Ahmed, K.; Rani, M.T.; Pradeep, K.S.; Al-Zahrani, F.A. Ultra-high negative dispersion compensating modified square shape photonic crystal fiber for optical broadband communication. *Alex. Eng. J.* **2022**, *61*, 2799–2806. [CrossRef]
47. An, G.; Li, S.; Zhang, W.; Fan, Z.; Bao, Y. A polarization filter of gold-filled photonic crystal fiber with regular triangular and rectangular lattices. *Opt. Commun.* **2014**, *331*, 316–319. [CrossRef]
48. Chao, C.-Y.; Guo, L.J. Design and optimization of micro ring resonators in biochemical sensing applications. *J. Lightwave Technol.* **2006**, *24*, 1395–1402. [CrossRef]
49. Liu, C.; Su, W.; Liu, Q.; Lu, X.; Wang, F.; Sun, T.; Chu, P.K. Symmetrical dual D-shape photonic crystal fibers for surface plasmon resonance sensing. *Opt. Express* **2018**, *26*, 9039–9049. [CrossRef] [PubMed]
50. Dasauni, P.; Mahapatra, M.; Saxena, R.; Kundu, S. Refractive index of blood is a potential qualitative indicator of haemoglobin disorder in humans. *J. Proteins Proteom.* **2018**, *9*, 1–21.
51. Rifat, A.A.; Hasan, M.R.; Ahmed, R.; Butt, H. Photonic crystal fiber-based plasmonic biosensor with external sensing approach. *J. Nanophoton.* **2017**, *12*, 012503. [CrossRef]
52. Chakma, S.; Khalek, M.A.; Paul, B.K.; Ahmed, K.; Hasan, M.R.; Bahar, A.N. Gold-coated photonic crystal fiber biosensor based on surface plasmon resonance: Design and analysis. *Sens. Bio-Sens. Res.* **2018**, *18*, 7–12. [CrossRef]
53. Amiri, I.S.; Alwi, S.A.K.; Raya, S.A.; Zainuddin, N.A.M.; Rohizat, N.S.; Rajan, M.M.; Zakaria, R. Graphene Oxide Effect on Improvement of Silver Surface Plasmon Resonance D-Shaped Optical Fiber Sensor. *J. Opt. Commun.* **2019**. [CrossRef]
54. De, M.; Markides, C.; Singh, V.K.; Themistos, C.; Rahman, B.M. Analysis of a Single Solid Core Flat Fiber Plasmonic Refractive Index Sensor. *Plasmonics* **2020**, *15*, 1429–1437. [CrossRef]
55. Zhang, Y.; Chen, H.; Wang, M.; Liu, Y.; Fan, X.; Chen, Q.; Wu, B. Simultaneous measurement of refractive index and temperature of seawater based on surface plasmon resonance in a dual D-type photonic crystal fiber. *Mater. Res. Express* **2021**, *8*, 085201. [CrossRef]
56. Bahloul, L.; Ferhat, M.L.; Haddouche, I.; Cherbi, L. A high-resolution refractive index sensor of partially cleaved PCF based on surface plasmon resonance. *J. Opt.* **2022**, *51*, 260–268. [CrossRef]
57. Kumar, D.; Sharma, M.; Singh, V. Surface Plasmon resonance implemented silver thin film PCF sensor with multiple–Hole microstructure for wide ranged refractive index detection. *Mater. Today Proc.* **2022**, *in press*. [CrossRef]
58. Meng, F.; Wang, H.; Fang, D. Research on D-Shape Open-Loop PCF Temperature Refractive Index Sensor Based on SPR Effect. *IEEE Photonics J.* **2022**, *14*, 5727605. [CrossRef]
59. Danlard, I.; Mensah, I.O.; Akowuah, E.K. Design and numerical analysis of a fractal cladding PCF-based plasmonic sensor for refractive index, temperature, and magnetic field. *Optik* **2022**, *258*, 168893. [CrossRef]

 micromachines MDPI

Article

Effect of the Deposition Time on the Structural, 3D Vertical Growth, and Electrical Conductivity Properties of Electrodeposited Anatase–Rutile Nanostructured Thin Films

Moisés do Amaral Amâncio [1], Yonny Romaguera-Barcelay [2], Robert Saraiva Matos [3], Marcelo Amanajás Pires [4], Ariamna María Dip Gandarilla [1], Marcus Valério Botelho do Nascimento [5], Francisco Xavier Nobre [5], Ştefan Ţălu [6,*], Henrique Duarte da Fonseca Filho [7] and Walter Ricardo Brito [1,*]

1 Department of Chemistry, Federal University of Amazonas-UFAM, Manaus 69067-005, AM, Brazil
2 Department of Physics, Federal University of Amazonas-UFAM, Manaus 69067-005, AM, Brazil
3 Graduate Program in Materials Science and Engineering, Federal University of Sergipe-UFS, São Cristóvão 49100-000, SE, Brazil
4 Brazilian Center for Research in Physics—CBPF, Rio de Janeiro 22290-180, RJ, Brazil
5 Federal Institute of Education, Science and Technology of Amazonas, Coari 69460-000, AM, Brazil
6 Directorate of Research, Development and Innovation Management (DMCDI), Technical University of Cluj-Napoca, 15 Constantin Daicoviciu St., 400020 Cluj-Napoca, Romania
7 Laboratory of Nanomaterials Synthesis and Nanoscopy, Department of Physics, Federal University of Amazonas, Manaus 69067-005, AM, Brazil
* Correspondence: stefan_ta@yahoo.com or stefan.talu@auto.utcluj.ro (Ş.Ţ.); wrbrito@ufam.edu.br (W.R.B.); Tel.: +55-92981019770 (W.R.B.)

Citation: do Amaral Amâncio, M.; Romaguera-Barcelay, Y.; Matos, R.S.; Pires, M.A.; Gandarilla, A.M.D.; do Nascimento, M.V.B.; Nobre, F.X.; Ţălu, Ş.; da Fonseca Filho, H.D.; Brito, W.R. Effect of the Deposition Time on the Structural, 3D Vertical Growth, and Electrical Conductivity Properties of Electrodeposited Anatase–Rutile Nanostructured Thin Films. *Micromachines* **2022**, *13*, 1361. https://doi.org/10.3390/mi13081361

Academic Editor: Chengyuan Dong

Received: 25 July 2022
Accepted: 17 August 2022
Published: 21 August 2022

Abstract: TiO_2 time-dependent electrodeposited thin films were synthesized using an electrophoretic apparatus. The XRD analysis revealed that the films could exhibit a crystalline structure composed of ~81% anatase and ~6% rutile after 10 s of deposition, with crystallite size of 15 nm. AFM 3D maps showed that the surfaces obtained between 2 and 10 s of deposition exhibit strong topographical irregularities with long-range and short-range correlations being observed in different surface regions, a trend also observed by the Minkowski functionals. The height-based ISO, as well as specific surface microtexture parameters, showed an overall decrease from 2 to 10 s of deposition, showing a subtle decrease in the vertical growth of the films. The surfaces were also mapped to have low spatial dominant frequencies, which is associated with the similar roughness profile of the films, despite the overall difference in vertical growth observed. The electrical conductivity measurements showed that despite the decrease in topographical roughness, the films acquired a thickness capable of making them increasingly insulating from 2 to 10 s of deposition. Thus, our results prove that the deposition time used during the electrophoretic experiment consistently affects the films' structure, morphology, and electrical conductivity.

Keywords: electrodeposition; ITO; morphology; thin films; TiO_2

1. Introduction

Titanium dioxide (TiO_2) nanopowders have received much interest due to their use in several technological applications, e.g., in industry and nanotechnology [1]. TiO_2 is used in many products, and its global demand growth is increasing rapidly along with its low price. It has various advantages, such as high surface tension, specific surface area, magnetic property, lower melting point, good thermal conductivity, and being environmentally friendly [2,3]. TiO_2 is a polymorphic compound with three mainly crystallographic phases: anatase, rutile, and brookite. They are different in their synthesis and properties, among which rutile and anatase are the most synthesized phases owing to their good thermodynamic characteristics and physical properties. Withal, anatase, and rutile titania exhibit an excellent refractive index, high dielectric constant, higher hiding power, and superior chemical stability [4,5].

TiO_2 thin films can be deposited using various methods, such as sol–gel dip and spin coating, pulsed laser deposition (PLD), spray pyrolysis, chemical vapor deposition (CVD), and sputtering [6,7]. Among these methods, chemical solution deposition is one of the more promising techniques because it provides higher composition control, lower processing temperatures, shorter fabrication time, and relatively low cost [8]. It is known that the coating deposition method can strongly influence the surface formed and affect properties, which justifies the choice of a method with high composition control.

The comprehension of the growth mechanisms and the study of structure and morphology as a function of the time deposition of the thin films are essential to preparing materials in a measured way for the desired properties. In this regard, studies based on the morphology of the thin films when the time deposition varies help to reveal the growth mechanism of these films [3]. For this purpose, scanning probe microscopy techniques have been employed. Atomic force microscopy (AFM) has been widely used for morphological characterization in real space and for determining thickness, roughness, and particle size in thin films [9]. Moreover, their 3D topographical maps can be used to analyze several other morphological parameters, e.g., spatial, hybrid, feature, functional and volumetric, and geometrical morphology shape and spatial frequencies, which can reveal the unique spatial patterns of the formed topography.

In the present work, we focus on the growth of nanostructured TiO_2 films on indium tin oxide (ITO) substrates obtained using the electrodeposition technique under different deposition times. This insight aims to analyze the influence of deposition time on the structural, morphological, and electrical conductivity properties of thin films. A correlation between time deposition, structure, surface roughness, growth morphology, and electrical properties of the TiO_2 films is established, which can help improve the fabrication processes of TiO_2-based devices. This study aims to obtain optimized conditions for electrodeposited TiO_2/ITO films at different times to enable photoelectrocatalytic and photovoltaic applications in future studies.

2. Materials and Methods

2.1. Chemicals and Materials

Potassium hexacyanoferrate (III), potassium chloride, ethyl alcohol (95%), metallic iodine, and titanium dioxide (P25) were purchased from Sigma-Aldrich (San Luis, MO, USA). Potassium hexacyanoferrate (II) trihydrate was obtained from Merck (Darmstadt, HD, Germany). Acetone, acetylacetone (99.5%), ethyl alcohol, and isopropyl alcohol were obtained from Synth (São Paulo, SP, Brazil). ITO/glass substrates (15 Ω/sq) were acquired from Lumtec (Taiwan, China).

2.2. Thin Films Electrophoretic Deposition

First, the electrolyte solution was prepared, for which 1.0 g of TiO_2 and 80 mL of ethyl alcohol were mixed in an Erlenmeyer flask. Then, 100 µL of acetylacetone was added to the system, which remained in magnetic stirring for 24 h. Afterward, 20 mg of metallic iodine was added to the solution, agitated for 30 min, and taken to an ultrasonic bath (Q335D, Quimis, São Paulo, SP, Brazil) for 20 min. The ITO electrodes (2 × 1 cm) were previously cleaned through successive sonication cycles in deionized water, acetone, and isopropyl alcohol for 15 min each stage [9]. Next, specific support for fixing the ITO electrodes was designed, using two connectors, and forming the anode and cathode with a separation of 2 mm. The electrode area to be coated was delimited to 1 cm^2, and the electrophoretic cell was filled with 10 mL of the electrolytic solution. For the electrodeposition process, the positive and negative poles were connected to the power supply (Agilent E3616A, Santa Clara, CA, USA) and applied a potential of 10 V. The formation of TiO_2 films occurs by electrostatic attraction of TiO_2 together with the negative pole electrode (anode), also known as TiO_2 anodization. Thus, TiO_2 films were obtained at different times (2, 4, 6, 8, and 10 s) in the TiO_2 solution. Subsequently, the films were treated thermally at 300 °C for 30 min in a EDG 3P-S muffle from EDG Equipment (São Paulo, SP, Brazil).

2.3. Film Characteristics Techniques

2.3.1. X-ray Diffraction Techniques

The X-ray diffraction (XRD) in grazing incident measurements was carried out with a LabX XRD-6000 apparatus from Shimadzu (Japan) and Cu–Kα radiation (λ = 1.5414 Å) source. The measurements were performed from 10 to 60 (2θ) in continuous (0.02/s) and step (0.02/10 s) modes, respectively. The XRD analysis was complemented with the Rietveld refinement technique with the GSAS 1.0 program.

2.3.2. AFM Measurements

An Innova AFM from Bruker (Billerica, MA, USA), operated on a taping mode with a scan rate of 0.5 Hz, was used to make the surface characterization. The samples were scanned in air and 40 ± 1% relative humidity over scanning areas of 5 × 5 μm^2, with a resolution of 256 × 256 pixels using a silicon cantilever (k = 40 N/m). The feedback control to obtain the best possible images was adapted to each surface and for all the applied scans. The analysis of the images was completed with the WSxM software (Madri, Spain), version 5.0, development 9.1 [10], and, through the images, it is possible to obtain the surface roughness parameters.

2.3.3. Surface Analysis

The complete analysis of the morphology of the films was based on the evaluation of 3D morphological parameters by the International Organization for Standardization (ISO) 25178-2: 2012 standard, whose parameters have their physical meaning largely described in references [11–15]. These parameters were obtained by the MountainsMap® 8.0 (Besançon, France) commercial software from Digital Surf [16]. In summary, we compute and evaluate several parameters, namely, height, feature, spatial, functional, hybrid, volume, and core Sk. Furthermore, we have determined the average power spectrum density (PSD) spectrum of fractal regions of the spectrum using linearized graphs obtained according to the mathematical theory explained by Jacobs et al. [17]. Additionally, the Hurst coefficients of the spectra were calculated using Equation (1), where α is the slope of the linearized curve obtained using the WSxM© 5.0 software [10].

$$Hc = \frac{\alpha - 2}{2} \tag{1}$$

Moreover, we know that 3D surface morphology at the micro/nanometric level is characterized by a series of descriptive parameters defined and quantified according to standard ISO 25178-2: 2012, as mentioned above. These descriptive shape parameters do not present a clear quantification of the nanostructured surface model for highlighting fineness characteristics such as arbitrary permutations of topographic positions of different heights. To study the topography more carefully, it was necessary to highlight them by applying integral geometry and morphological descriptors, known as Minkowski functionals (MFs), which characterize both the connectivity (topology) and the content and shape (geometry) of spatial models [18–20]. For this purpose, we consider the MFs defined for a convex, compact set $K \subset R^3$ via Steiner's formula. So, let $K \oplus B_r$ be the dilation of the set K by a closed ball of radius r centered on the origin. The volume $V^{(3)}$ (for dimension d = 3) of $K \oplus B_r$ can be written as a polynomial function of r as follows [19,20]:

$$V^{(3)}(K \oplus B_r) = \sum_{k=0}^{3} \binom{3}{k} W_k^{(3)}(K) r^k \tag{2}$$

where ${W_k}^{(3)}$ is the kth Minkowski functional.

The MFs can be expressed using the common descriptors volume V, surface area S, mean breadth B, and Euler–Poincaré characteristic χ by the following relations [18,19]:

$$W_0^{(3)}(K) = V(K); W_1^{(3)}(K) = \frac{1}{3}S(K); W_2^{(3)}(K) = \frac{2}{3}\pi \cdot B(K); W_3^{(3)}(K) = \frac{4}{3}\pi \cdot \chi(K) \quad (3)$$

2.3.4. Electrochemical Measurements

The electrochemical measurements were collected in an AUTOLAB PGSTAT 128N (Metrohm Autolab, Utrecht, The Netherlands), controlled with NOVA version 2.1.5 electrochemical analysis software. The electrochemical cell was constituted by three electrodes, using TiO_2/ITO as the working electrode, Ag/AgCl (sat. KCl) as the reference electrode, and platinum as the auxiliary electrode. The experiments were performed at 22 ± 1 °C, without stirring. The films were characterized by electrochemical impedance spectroscopy (EIS) in a 5 mmol L^{-1} $[Fe(CN)6]^{4-/3-}$ + 0.1 mol L^{-1} KCl solution at 0.2 V, varying the frequency 10 points per decade in the range from 10 kHz to 100 MHz.

3. Results and Discussion

Figure 1 shows the XRD pattern of semiconductor titanium oxide deposited using the immersion method at 300 °C for 2, 4, 6, 8, and 10 s. The diffraction peaks at 25.82°, 38.31°, 41.82°, 48.55°, 55.71°, and 63.34° corresponding to the planes (101), (004), (112), (200), (105), and (215) were observed, indicating the formation of TiO_2 anatase phase being ~79% of the main phase in all samples [19–21]. These signals match well with the peaks previously reported for TiO_2 in inorganic crystal structure database (ICSD) card No. 9852 and ICSD card No. 24,276 [21,22], which is ~7% of the ITO phase. Other peaks appear in the XRD pattern at 27.97°, 35.72°, and 54.54°, and can be assigned to the planes (110), (101), and (211), revealing the presence of TiO_2 rutile phase (TiO_2, ICSD card No. 9161) [23], which is 14% of the other phases. The peaks of indium oxide and tin substrate (indicated as ITO in Figure 1e) were also registered. New phases are not observed in the different immersion times.

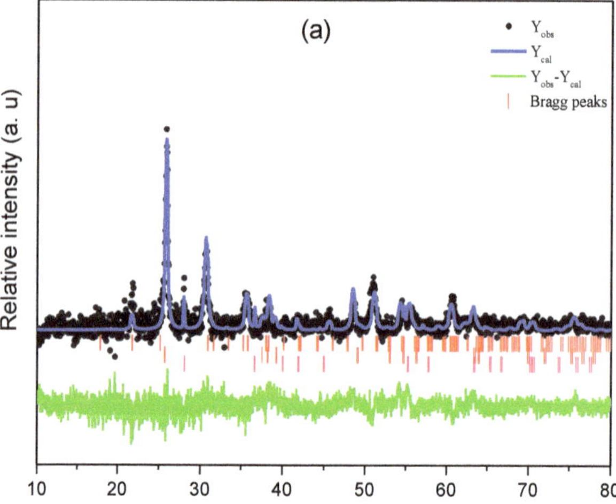

Figure 1. *Cont.*

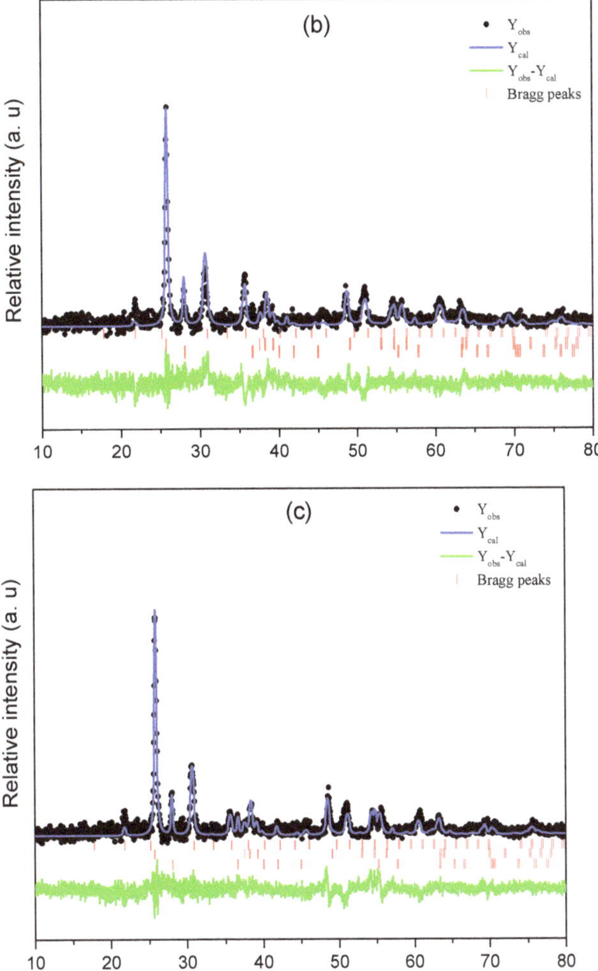

Figure 1. *Cont.*

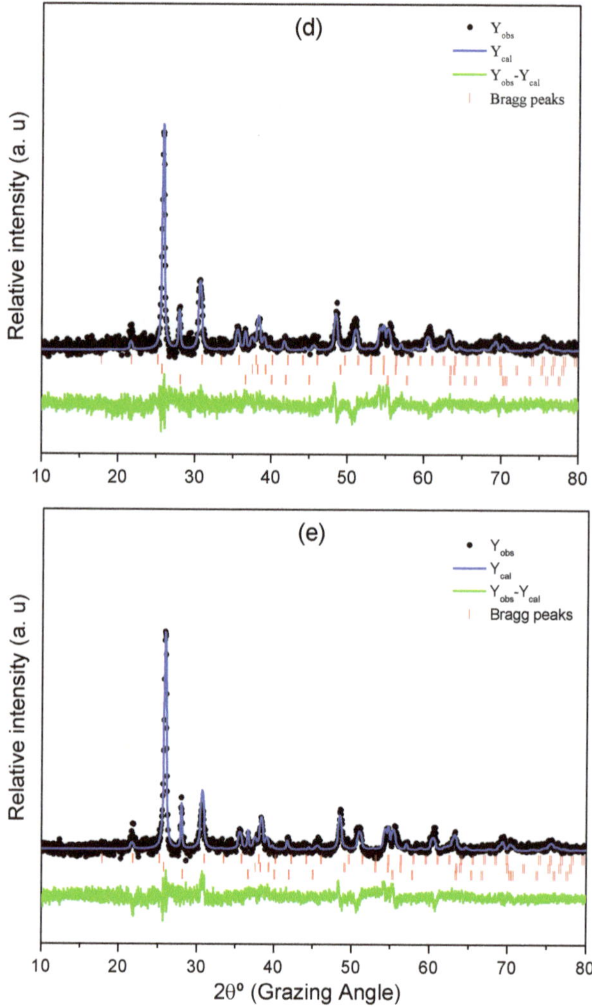

Figure 1. XRD spectra of TiO_2 thin films at different immersion times: (**a**) 2 s, (**b**) 4 s, (**c**) 6 s, (**d**) 8 s, and (**e**) 10 s. Experimental (Yobs) and theoretical (Yobs) data, in which the residual line (Yobs–Ycal) and Bragg peaks for all samples.

XRD spectra of TiO_2 thin films were refined and used simultaneously for both structures, anatase (ICSD No. 9852) [21,22] and rutile (ICSD No. 9161) [23], assigned to a tetragonal structure, with a space group *I41/amd* (141) and space group *P42/mnm* (136), respectively. The phase composition analysis revealed that anatase was the major phase at 81% ± 4% in 10 s electrodeposition time, while having lower rutile phase of 6.1%.

Table S1 (Supplementary Information) shows the lattice parameter of TiO_2 thin films obtained of the refined structure for anatase and rutile in function electrodeposition time. The results verified the correlation between the experimental and theoretical intensities for both samples (Figure 1a,e), as observed in the residual line.

The crystallite size (*D*) for the prepared samples was determined by measuring the broadening of the most intense (best defined) peak of the phase in a diffraction pattern associated with a specific planar reflection within the crystal unit cell according to the Debye–Scherrer equation as follows [24]:

$$D = \frac{k\lambda}{\beta \cdot \cos\theta} \quad (4)$$

where k is a correction factor account for particle shapes (in this work, $k = 9$ was used), λ is the wavelength of Cu target = 1.5406 Å, β is the full width at half maximum (FWHM) of the most intense diffraction peak plane, and θ is the Bragg's angle. The crystalline size calculated for TiO_2 films with different immersion times evidence that anatase structures have a size of 15 nm for 10 s of immersion compared to rutile structures. Nevertheless, both structures have similar crystalline values in the range from 4 to 6 s, shown in Table S1.

3.1. Morphology Analysis

Figure 2 provides the first insights into the topographic properties of the electrodeposited TiO_2 thin films. The surfaces display an aspect randomly rough on a global scale with some spatial domains, manifesting long-range correlations and other regions exhibiting short-range correlations. Also evident is the hierarchical character of the surface roughness, as several asperities are composed of more asperities that, in turn, are formed with more asperities. Along with the visual information conveyed in Figure 2, we can obtain a quantitative assessment of the topography with the set of height-based parameters, which are provided in Table 1.

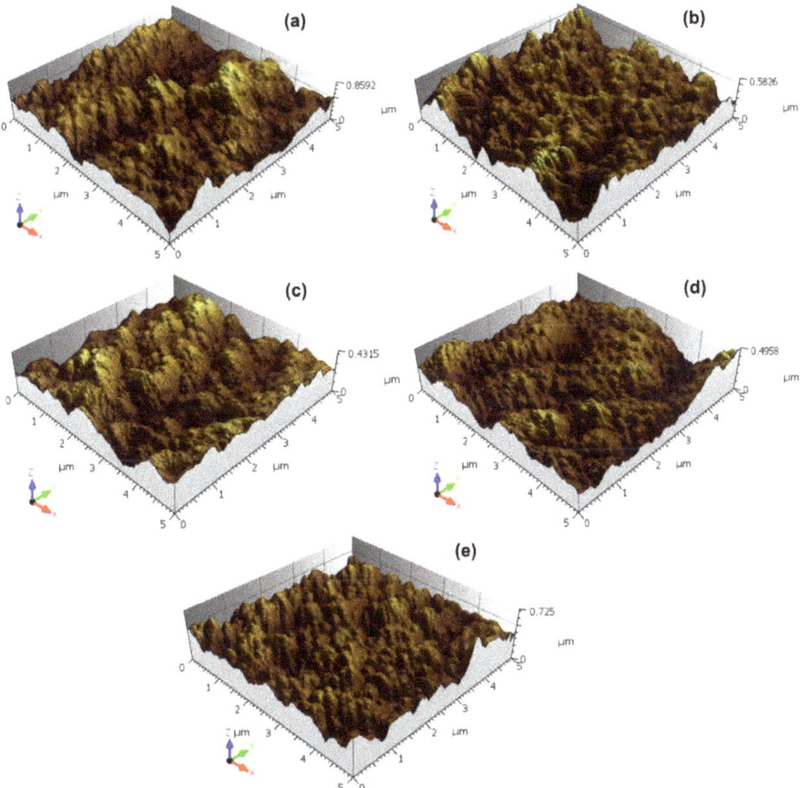

Figure 2. Three-dimensional AFM images of the TiO_2 thin films with different electrodeposition times: (**a**) 2 s, (**b**) 4 s, (**c**) 6 s, (**d**) 8 s and (**e**) 10 s.

Table 1. Surface parameters of the thin films, according to ISO 25178-2:2012.

Par [µm]	Electrodeposition Time				
	2 s	4 s	6 s	8 s	10 s
Sq	0.122 ± 0.019	0.108 ± 0.023	0.068 ± 0.010	0.068 ± 0.006	0.091 ± 0.008
Sp	0.411 ± 0.049	0.358 ± 0.085	0.233 ± 0.058	0.267 ± 0.023	0.336 ± 0.076
Sv	0.403 ± 0.046	0.471 ± 0.148	0.234 ± 0.020	0.239 ± 0.014	0.353 ± 0.067
Sz	0.814 ± 0.088	0.829 ± 0.233	0.467 ± 0.070	0.506 ± 0.031	0.688 ± 0.137

Table 1 depicts the temporal values for the root mean square height (Sq), maximum peak height (Sp), maximum pit height (Sv), and maximum height (Sz). In the films obtained with electrodeposition time of 4 to 10 s, an overall consonance between the trends of increase/decrease is noted, related to extreme heights, despite some fluctuations. The concomitant minimum value further emphasizes this at the same intermediate time (t = 6 s).

3.2. Advanced Stereometry Evaluation

Table 2 presents the values for the spatial and hybrid parameters [11]. The results did not show significant differences in the autocorrelation length (Sal) for all films, without evidence of considerable increment or decrement in the relative predominance of correlated spatial events. The texture aspect ratio (Str) was another quantity that did not show significant changes over time, suggesting statistical robustness in the magnitude of the surface texture anisotropy from t = 2 s to t = 10 s. Similarly, the texture direction (Std) did not show noticeable changes over time, indicating that the dominant orientation of the surface texture anisotropy persists relatively constant from t = 2 s to t = 10 s.

Table 2. Spatial and hybrid parameters of the thin films, in accordance with ISO 25178-2:2012.

Par	Unit	Electrodeposition Time				
		2 s	4 s	6 s	8 s	10 s
			Spatial			
Sal *	[µm]	0.432 ± 0.044	0.353 ± 0.037	0.383 ± 0.021	0.375 ± 0.050	0.338 ± 0.049
Str *	[-]	0.472 ± 0.011	0.427 ± 0.136	0.394 ± 0.191	0.603 ± 0.069	0.613 ± 0.126
Std *	[°]	123.94 ± 67.30	51.75 ± 72.72	52.00 ± 69.28	99.87 ± 77.46	118.44 ± 63.85
			Hybrid			
Sdq	[-]	961.95 ± 54.25	978.06 ± 114.74	495.91 ± 31.63	672.96 ± 141.40	878.85 ± 80.38
Sdr	[%]	75.33 ± 3.09	75.75 ± 6.29	42.00 ± 2.37	55.17 ± 11.35	70.08 ± 5.40

* Samples without significant difference, one-way ANOVA and Tukey's test ($p < 0.05$).

Now let us focus on the hybrid parameters, which are quantities that have as input the content of the three spatial directions. The values for the root mean square gradient (Sdq) (that is zero for a flat surface) reveal that the samples for t = 6 s present the smallest local steepness. This local property of the surface is corroborated with the interfacial area ratio (Sdr) results, achieving the smallest value for t = 6 s. Such a qualitative agreement is also noted for other time values of Sdq and Sdr, signaling that the local steepness and the relative interfacial surface area incrementally present a coupled response over time in the current set of experiments.

Table 3 conveys information about the temporal behavior of the feature parameters. The density of peaks (Spd) did not show substantial changes over time, suggesting statistical robustness in the number of points (per unit area) for potential contact with other surfaces. On the other hand, the arithmetic means peak curvature (Spc) exhibited strong changes over time where the maximum value occurred for t = 8 s, indicating that the curvature of peaks becomes larger (peaks more rounded) at such an instant. The ten-point height of surface (S10z) and five-point peak height (S5p) attains the minimum value concomitantly at t = 6 s. In contrast, the five-point pit depth (S5v) reaches the minimum value at t = 8 s, suggesting a close temporal behavior between peaks and valleys in the samples. While the mean hill area (Sha) manifested substantial changes, the mean dale area (Sda), mean dale volume (Sdv), and mean hill volume (Shv) presented values without statistical difference. This indicates that the process of formation of the thin films comprised a coupled evolution of the Sda, Sdv, and Shv.

Table 3. Feature parameters of the thin films, in accordance with ISO 25178-2:2012.

Par	Unit	Electrodeposition Time				
		2 s	4 s	6 s	8 s	10 s
Spd *	[1/μm²]	2.290 ± 0.479	2.620 ± 0.778	1.990 ± 0.401	3.700 ± 1.261	2.940 ± 0.254
Spc	[1/μm]	32561 ± 7825	30270 ± 9318	18006 ± 1691	36391 ± 11325	18643 ± 1490
S10z	[μm]	0.700 ± 0.088	0.705 ± 0.175	0.381 ± 0.050	0.413 ± 0.050	0.579 ± 0.094
S5p	[μm]	0.350 ± 0.041	0.319 ± 0.068	0.190 ± 0.030	0.225 ± 0.030	0.265 ± 0.045
S5v	[μm]	0.350 ± 0.050	0.386 ± 0.107	0.191 ± 0.021	0.188 ± 0.023	0.315 ± 0.054
Sda *	[μm²]	0.727 ± 0.225	0.508 ± 0.095	0.769 ± 0.114	0.721 ± 0.313	0.516 ± 0.033
Sha	[μm²]	0.471 ± 0.083	0.410 ± 0.093	0.575 ± 0.141	0.317 ± 0.112	0.365 ± 0.027
Sdv *	[μm³]	2.532 ± 0.535	2.709 ± 0.302	2.077 ± 0.684	3.293 ± 1.816	2.387 ± 1.065
Shv *	[μm³]	4.404 ± 1.249	4.063 ± 1.827	4.482 ± 2.068	1.565 ± 0.758	4.044 ± 1.313

* Samples without significant difference, one-way ANOVA and Tukey's test ($p < 0.05$).

The panels in Figure 3 reveal that the height histograms have a bell-like shape (with a basis on the y-axis). In turn, the red curve overlaid on the histogram is a type of Abbott–Firestone curve [25] that is a suitable statistical method for characterizing surfaces. Such a curve is a cumulative probability density function that, in practice, is estimated from cumulative sums over the height histogram in a way that its minimum value (0%) is obtained for the highest peak, and its maximum is attained for the deepest valley (100%). The results show that the samples of TiO_2 thin films produce Abbott–Firestone curves with a quite smooth S shape (looking from the y-axis). This suggests that for successive depths (from the highest peak), there is a gradual increase in the percentage of the material traversed in relation to the area spanned. The panels {b,c, I–f, h,i, l,m, o,p} provide a statistical interpretation for a plethora of functional parameters [26], which will be discussed further.

Table 4 unveils that the areal material ratio (Smr), peak material portion (Smr1), and valley material portion (Smr2) present variations without statistical difference. This means that the percentage of material that constitutes the peak/valley patterns remains stable over time, despite some fluctuations. It is an obvious general congruence between the patterns of increase/decrease of the inverse areal material ratio (Smc), peak extreme height (Sxp), core roughness depth (Sk), reduced peak height (Spk), and the reduced valley depth (Svk), although there are punctual discrepancies. Similarly, broadly correlated the successive tendencies of growths/declines of dale void volume (Vvv), core void volume (Vvc), peak material volume (Vmp), and core material volume (Vmc). In summary, the results for several functional parameters associated with superficial strata (Smc, Sxp, Sk, Spk, and Svk) and volume (Vvv, Vvc, Vmp, and Vmc) show a correspondence between the temporal variations in distinct portions of the Abbott–Firestone curve obtained from the samples. This points out the overall compatibility between the changes across different slices of the surface morphology.

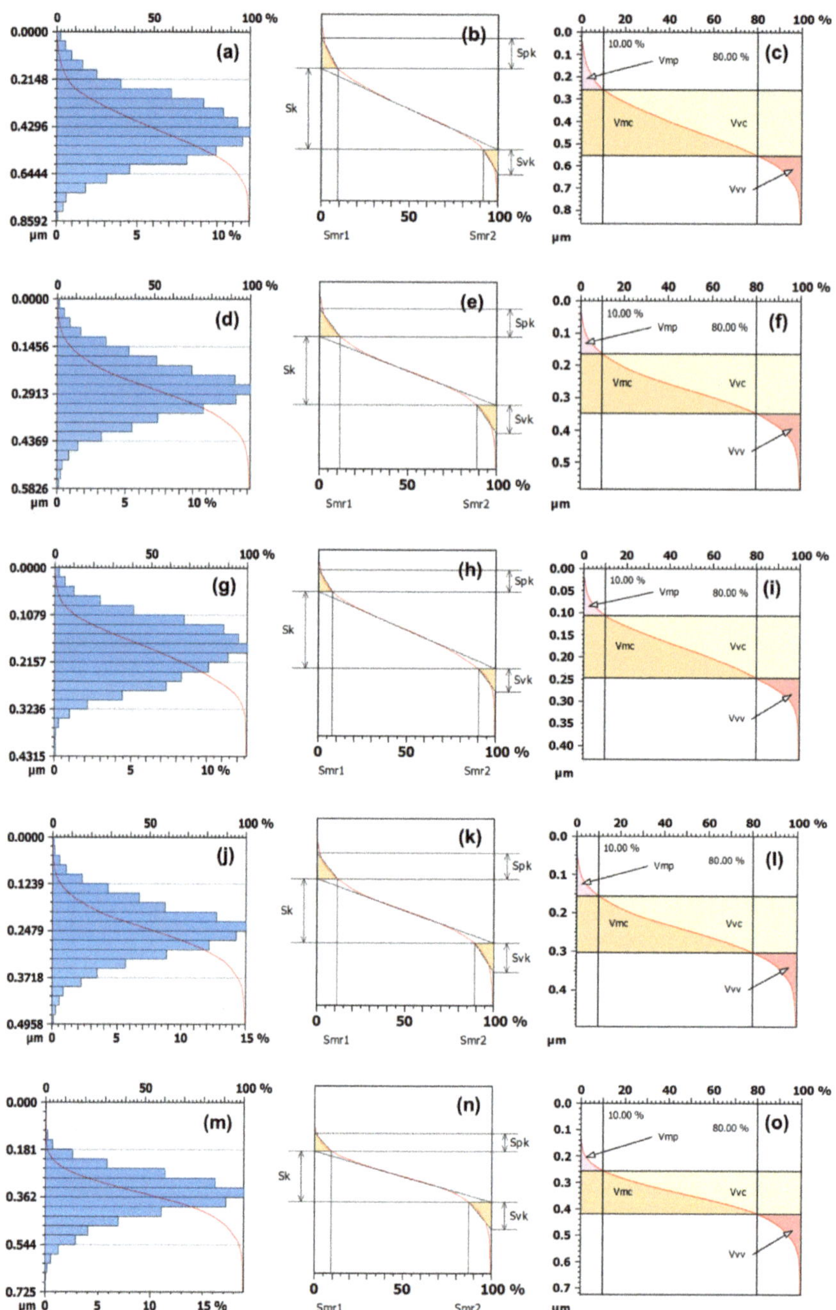

Figure 3. Most relevant Sk and volume parameters representation of (**a**–**c**) 2 s, (**d**–**f**) 4 s, (**g**–**i**) 6 s, (**j**–**m**) 8 s, and (**n**,**o**) 10 s. The horizontal axis in (**c**,**f**) is related to the percentage of occupied volume. The vertical axis is related to depth; 10% and 80% are used to divide the reduced peaks and reduced valleys from the core surface [11].

Table 4. Functional parameters of the thin films, according to ISO 25178-2:2012.

Par	Unit	Electrodeposition Time				
		2 s	4 s	6 s	8 s	10 s
				Functional		
Smr *	[%]	0.002 ± 0.001	0.002 ± 0.001	0.002 ± 0.001	0.003 ± 0.002	0.003 ± 0
Smc	[μm]	0.155 ± 0.027	0.134 ± 0.027	0.087 ± 0.013	0.086 ± 0.006	0.111 ± 0.002
Sxp	[μm]	0.239 ± 0.032	0.237 ± 0.063	0.133 ± 0.019	0.132 ± 0.016	0.189 ± 0.016
Sk	[μm]	0.289 ± 0.054	0.231 ± 0.027	0.175 ± 0.024	0.166 ± 0.012	0.213 ± 0.011
Spk	[μm]	0.131 ± 0.019	0.121 ± 0.036	0.063 ± 0.013	0.076 ± 0.008	0.096 ± 0.038
Svk	[μm]	0.127 ± 0.018	0.147 ± 0.052	0.063 ± 0.008	0.070 ± 0.008	0.107 ± 0.016
Smr1 *	[%]	11.837 ± 2.197	12.263 ± 1.255	9.786 ± 1.317	10.899 ± 0.298	9.952 ± 0.361
Smr2 *	[%]	89.087 ± 2.088	87.854 ± 1.082	89.750 ± 0.913	89.708 ± 0.864	87.83 ± 0.929
				Volume		
Vmp	[$\mu m^3/\mu m^2$]	6.265 ± 0.823	5.618 ± 1.510	3.205 ± 0.616	3.761 ± 0.462	4.798 ± 1.898
Vmc	[$\mu m^3/\mu m^2$]	105.80 ± 19.58	87.56 ± 13.97	62.73 ± 10.60	59.32 ± 5.25	78.29 ± 4.29
Vvc	[$\mu m^3/\mu m^2$]	147.50 ± 25.924	125.22 ± 24.05	82.89 ± 12.28	81.86 ± 5.82	102.37 ± 6.58
Vvv	[$\mu m^3/\mu m^2$]	14.241 ± 2.300	14.84 ± 4.327	7.504 ± 0.972	7.847 ± 0.837	11.49 ± 1.027

* Samples without significant difference, one-way ANOVA and Tukey's test ($p < 0.05$).

3.3. Power Spectrum Density of the TiO_2 Thin Films Nanotexture

PSD has been an excellent mathematical tool for evaluating patterns on surfaces with different dominant wavelengths [27–29]. The linearized graphs of the obtained average PSD spectrum are shown in Figure 4, whose Hurst coefficient values are exposed in Table 5. Although the films exhibited different morphologies, their roughness has similar special frequencies because the spectra are in similar regions. The statistical analysis found no difference between H values, although its mean value has decreased from 2 to 8 s. This suggests that the surface nanotexture does not changed as the deposition time increased, in accordance with Sal, Str, and Std parameters. Due to the film's uniform deposition, the nanotexture does not showed statistically different patterns of dominant spatial frequency (similar dominant wavelength). This nanotexture characteristic can also favor some similar physical properties, in addition to their homogeneity, as observed by Barcelay et al. [15].

Table 5. Hurst coefficient (H) of the TiO_2 thin films with different electrodeposition times: 2, 4, 6, 8, and 10 s.

Par	Electrodeposition Time				
	2 s	4 s	6 s	8 s	10 s
H *	0.53 ± 0.04	0.51 ± 0.08	0.48 ± 0.07	0.41 ± 0.16	0.59 ± 0.04

* Samples without significant difference, one-way ANOVA and Tukey's test ($p < 0.05$).

3.4. Minkowski Functionals

The MFs of AFM images were calculated and averaged over various positions on each surface. The resulting graphs for Minkowski volume V, Minkowski boundary S, and Minkowski connectivity χ are plotted in Figure 5. As can be seen, three descriptors (Minkowski volume, Minkowski limit, and Minkowski connectivity) show a different graphical evolution, which highlights a distinct 3D morphology of the surface texture with different types of nanopatterned templates. For the Minkowski volume V (Figure 5a,b), the lowest values are for samples with t = 2 s, and the highest values correspond to samples with t = 10 s. The Minkowski boundary S (Figure 5c,d) presents a maximum peak in the samples with t = 10 s. Furthermore, the Minkowski connectivity χ (Figure 5e,f) describes the topological structure of the surfaces' patterns. All graphs have an oscillated form with values dispersed in a wide range of heights. There is a maximum peak for samples with t = 8 s, while the minimum value is associated with samples with t = 10 s.

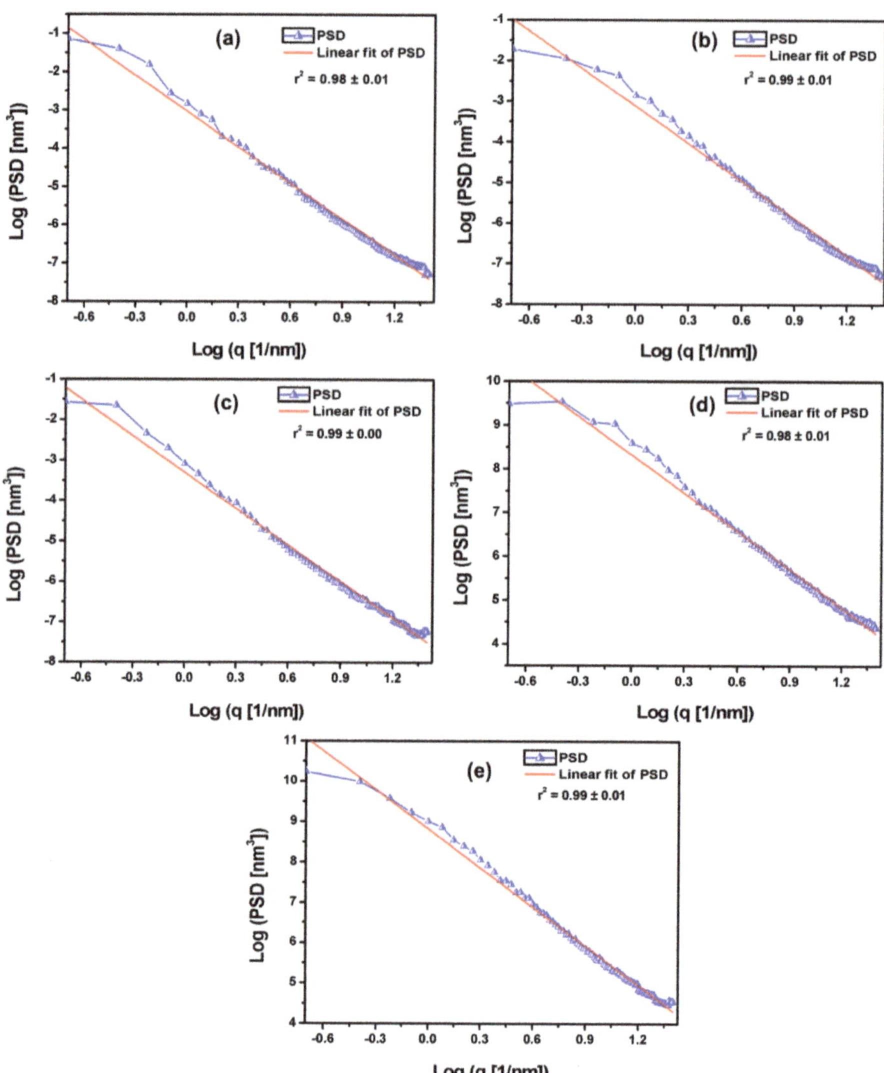

Figure 4. Average power spectrum density of the TiO_2 thin films with different electrodeposition times: (**a**) 2 s, (**b**) 4 s, (**c**) 6 s, (**d**) 8 s and (**e**) 10 s.

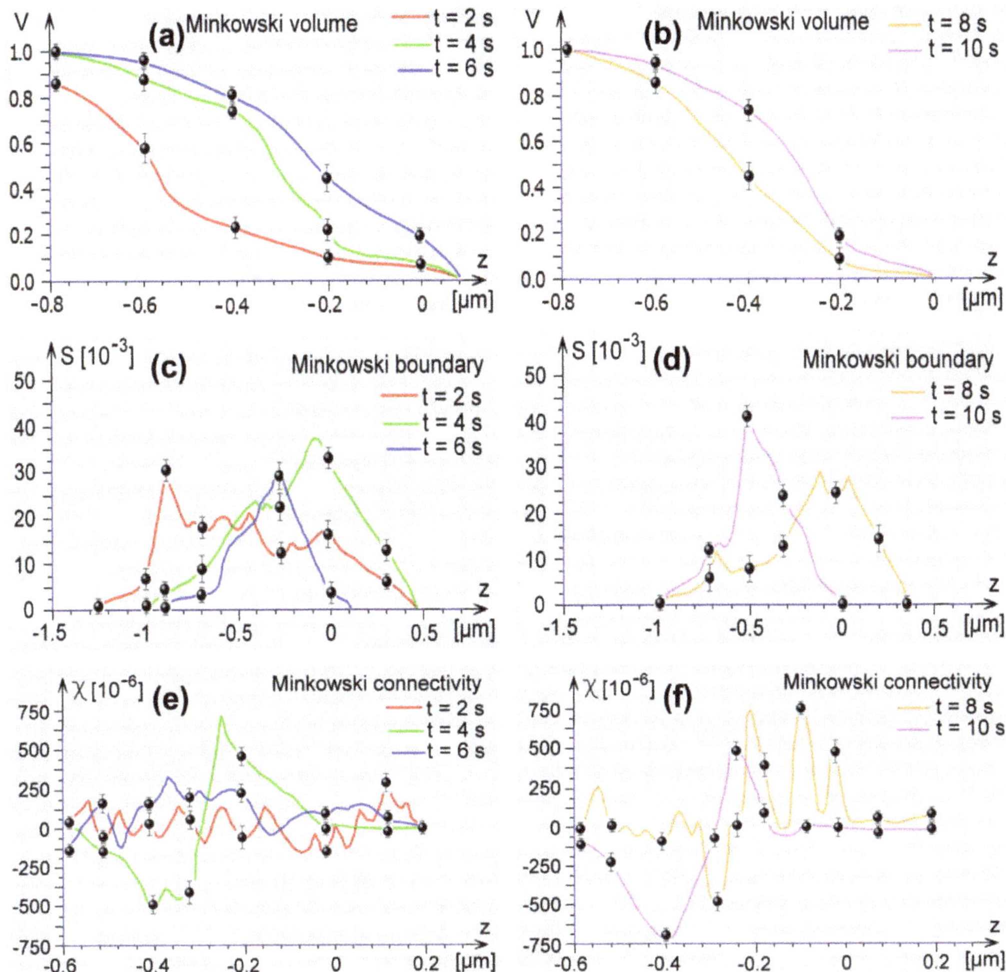

Figure 5. Minkowski volume of the thin films for: (**a**) 2 s, 4 s, and 6 s; (**b**) 8 s, and 10 s; Minkowski boundary of the thin films for: (**c**) 2 s, 4 s, and 6 s; (**d**) 8 s, and 10 s; Minkowski connectivity of the thin films for: (**e**) 2 s, 4 s, and 6 s; (**f**) 8 s, and 10 s.

3.5. Electrochemical Impedance Spectroscopy of the TiO_2 Films

The results of EIS are represented as Nyquist graphs (Figure 6), which include a semicircle in high-frequency ranges representing the charge transfer resistance (Rct) and a linear part at lower frequencies representing the diffusion process. The resistance values (Rct) were calculated using the Randles equivalent circuit (inset Figure 6), where Rs is the solution and connectors resistance, Zw is the Warburg impedance that corresponds to mass transfer of the redox species to the electrode, and constant phase element (CPE) is the impedance of the electrode/solution interface and is associated with the formation of the double electrical layer [30].

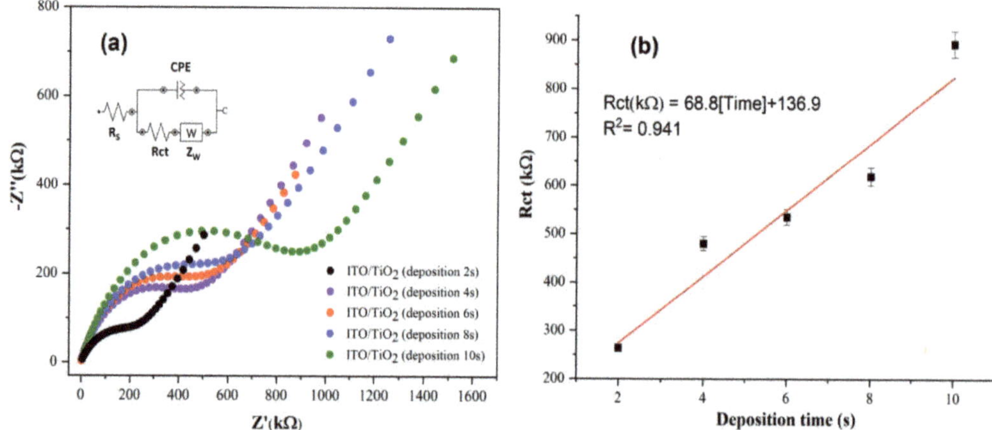

Figure 6. Electrochemical impedance spectroscopy in $[Fe(CN)_6]^{4-/3-}$ + KCl solution. (**a**) Nyquist diagrams of ITO/TiO_2 with the application of different deposition times, and (**b**) calibration curve (Plots of Rct versus deposition time).

As can be observed (Figure 6a), the semicircle at high frequencies increased with the deposition time due to the formation of a thicker TiO_2 layer, attributing insulating properties to the film. The calibration curve was obtained by plotting resistance (Rct) (kΩ) versus deposition time (Figure 6b), where a linear behavior is observed. The equation of regression was defined as Rct (kΩ) = 68.8 [deposition time] + 136.9, with $R^2 = 0.941$.

4. Conclusions

In summary, anatase–rutile time-dependent electrodeposited thin films were successfully obtained in this work. The films present a tetragonal crystalline structure ascribed to the anatase phase (81%), but also exhibit small amounts (~6%) of rutile after 10 s of deposition. The crystallite size of all films was found to be around 15 nm. The long- and short-range correlations found along the film's surface are assigned to the main phenomena that characterize the strong topographical irregularities of the films. Despite this, the vertical growth decreases as the deposition time increases, showing that the surface of the films becomes smoother. This trend was confirmed by the analysis of Minkowski functionals. PSD results indicate that all films are dominated by dominant low spatial frequencies, which can be attributed to similar surface microtexture, although the surfaces have different morphologies. The electrical conductivity tests showed that the films become more insulating as the deposition time increases, which notably occurs thanks to the increase in the thickness of the oxide layer that is produced on the ITO substrate. Our findings prove that the structure, morphology, and electrical conductivity of anatase–rutile films could be controlled by the deposition time employed in the electrophoretic essay, showing that this physical parameter plays an important role in the film's physical properties.

Supplementary Materials: The following are available online at https://www.mdpi.com/article/10.3390/mi13081361/s1, Table S1: Calculated Rietveld refinement results for XRD measurements performed on electrodeposited TiO_2 thin films.

Author Contributions: M.d.A.A. and Y.R.-B.: project coordination, methodology, and administration. Y.R.-B., R.S.M. and M.A.P.: conceptualization, methodology, and data collecting/analysis. A.M.D.G.: investigation and data curation. M.V.B.d.N. and F.X.N.: original draft preparation, data analysis, and visualization. Ş.Ţ.: investigation, writing, reviewing, and funding acquisition. H.D.d.F.F., W.R.B. and R.S.M.: measurement and data interpretation, resources, supervision, and software. All authors have read and agreed to the published version of the manuscript.

Funding: The authors would like to thank for the financial support to Coordenação de Aperfeiçoamento de Pessoal de Nível Superior (CAPES) and the Analytical Center of the Federal University of Amazonas (UFAM). We also would like to thank the Laboratório de Síntese de Nanomateriais e Nanoscopia (LSNN) associated to the Sistema Nacional de Laboratórios em Nanotecnologias (SisNANO)-Processo CNPq 442601/2019-0.

Data Availability Statement: The processed data required to reproduce these findings are available from the corresponding authors.

Conflicts of Interest: The authors declare no conflict of interest.

References

1. Haider, A.J.; Jameel, Z.N.; Al-Hussaini, I.H.M. Review on: Titanium Dioxide Applications. *Energy Procedia* **2019**, *157*, 17–29. [CrossRef]
2. Hoseinzadeh, T.; Solaymani, S.; Kulesza, S.; Achour, A.; Ghorannevis, Z.; Ţălu, Ş.; Bramowicz, M.; Ghoranneviss, M.; Rezaee, S.; Boochani, A.; et al. Microstructures, fractal geometry and dye-sensitized solar cells performance of CdS/TiO_2 nanostructures. *J. Electroanal. Chem.* **2018**, *830-831*, 80–87. [CrossRef]
3. Zare, M.; Solaymani, S.; Shafiekhani, A.; Kulesza, S.; Ţălu, Ş.; Bramowicz, M. Evolution of rough-surface geometry and crystalline structures of aligned TiO_2 nanotubes for photoelectrochemical water splitting. *Sci. Rep.* **2018**, *8*, 10870. [CrossRef] [PubMed]
4. Ryu, Y.B.; Lee, M.S.; Jeong, E.D.; Kim, H.G.; Jung, W.Y.; Baek, S.H.; Lee, G.D.; Park, S.S.; Hong, S.S. Hydrothermal Synthesis of Titanium Dioxides from Peroxotitanate Solution Using Different Amine Group-Containing Organics and Their Photocatalytic Activity. *Catal. Today* **2007**, *124*, 88–93. [CrossRef]
5. Shakoury, R.; Arman, A.; Ţălu, Ş.; Dastan, D.; Luna, C.; Rezaee, S. Stereometric analysis of TiO_2 thin films deposited by electron beam ion assisted. *Opt. Quantum Electron.* **2020**, *52*, 270. [CrossRef]
6. Korpi, A.R.G.; Rezaee, S.; Luna, C.; Ţălu, Ş.; Arman, A.; Ahmadpourian, A. Influence of the oxygen partial pressure on the growth and optical properties of RF-sputtered anatase TiO_2 thin films. *Results Phys.* **2017**, *7*, 3349–3352. [CrossRef]
7. Silva-Moraes, M.O.; Garcia-Basabe, Y.; de Souza, R.F.B.; Mota, A.J.; Passos, R.R.; Galante, D.; Fonseca Filho, H.D.; Romaguera-Barcelay, Y.; Rocco, M.L.M.; Brito, W.R. Geometry-Dependent DNA-TiO_2 Immobilization Mechanism: A Spectroscopic Approach. *Spectrochim. Acta Part A Mol. Biomol. Spectrosc.* **2018**, *199*, 349–355. [CrossRef]
8. Ţălu, Ş.; Achour, A.; Solaymani, S.; Nikpasand, K.; Dalouji, V.; Sari, A.; Rezaee, S.; Nezafat, N.B. Micromorphology analysis of TiO_2 thin films by atomic force microscopy images: The influence of postannealing. *Microsc. Res. Tech.* **2020**, *83*, 457–463. [CrossRef]
9. Franco, L.A.; Sinatora, A. 3D Surface Parameters (ISO 25178-2): Actual Meaning of Spk and Its Relationship to Vmp. *Precis. Eng.* **2015**, *40*, 106–111. [CrossRef]
10. Horcas, I.; Fernández, R.; Gómez-Rodríguez, J.M.; Colchero, J.; Gómez-Herrero, J.; Baro, A.M. WSXM: A Software for Scanning Probe Microscopy and a Tool for Nanotechnology. *Rev. Sci. Instrum.* **2007**, *78*, 013705. [CrossRef]
11. Blateyron, F. *Characterisation of Areal Surface Texture*; Leach, R., Ed.; Springer: Berlin/Heidelberg, Germany, 2013; Volume 9783642364, ISBN 978-3-642-36457-0.
12. Matos, R.S.; Pinto, E.P.; Ramos, G.Q.; de Albuquerque, M.D.F.; da Fonseca Filho, H.D. Stereometric Characterization of Kefir Microbial Films Associated with Maytenus Rigida Extract. *Microsc. Res. Tech.* **2020**, *83*, 1401–1410. [CrossRef] [PubMed]
13. Barcelay, Y.R.; Moreira, J.A.G.; de Jesus Monteiro Almeida, A.; Brito, W.R.; Matos, R.S.; da Fonseca Filho, H.D. Nanoscale Stereometric Evaluation of $BiZn_{0.5}Ti_{0.5}O_3$ Thin Films Grown by RF Magnetron Sputtering. *Mater. Lett.* **2020**, *279*, 128477. [CrossRef]
14. Ţălu, Ş. *Micro and nanoscale characterization of three dimensional surfaces. Basics and Applications*; Napoca Star Publishing House: Cluj-Napoca, Romania, 2015; pp. 199–210.
15. Jacobs, T.D.B.; Junge, T.; Pastewka, L. Quantitative Characterization of Surface Topography Using Spectral Analysis. *Surf. Topogr. Metrol. Prop.* **2017**, *5*, 013001. [CrossRef]
16. Mountains®8 Digital Surf Mountains®8 Digital Surf Headquarters, Besançon, France. 2020. Available online: https://www.digitalsurf.com/ (accessed on 24 July 2022).
17. Sevink, G.J.A. Mathematical Description of Nanostructures with Minkowski Functionals. In *Nanostructured Soft Matter*; Springer: Dordrecht, The Netherlands, 2007; pp. 269–299.
18. Mecke, K.R.; Stoyan, D. (Eds.) Statistical Physics and Spatial Statistics. In *Lecture Notes in Physics*; Springer: Berlin/Heidelberg, Germany, 2000; Volume 554, ISBN 978-3-540-67750-5.
19. Pal, S.; Laera, A.M.; Licciulli, A.; Catalano, M.; Taurino, A. Biphase TiO_2 Microspheres with Enhanced Photocatalytic Activity. *Ind. Eng. Chem. Res.* **2014**, *53*, 7931–7938. [CrossRef]
20. Du, Y.; Niu, X.; Li, W.; An, J.; Liu, Y.; Chen, Y.; Wang, P.; Yang, X.; Feng, Q. Microwave-Assisted Synthesis of High-Energy Faceted TiO_2 Nanocrystals Derived from Exfoliated Porous Metatitanic Acid Nanosheets with Improved Photocatalytic and Photovoltaic Performance. *Materials* **2019**, *12*, 3614. [CrossRef]
21. Zhang, H.; Banfield, J.F. Understanding Polymorphic Phase Transformation Behavior during Growth of Nanocrystalline Aggregates: Insights from TiO_2. *J. Phys. Chem. B* **2000**, *104*, 3481–3487. [CrossRef]
22. Schossberger, F. Ueber Die Umwandlung Des Titandioxyds. *Z. Krist.* **1942**, *104*, 358–374.

23. Horn, M.; Schwerdtfeger, C.F.; Meagher, E.P. Refinement of the Structure of Anatase at Several Temperatures. *Z. Krist. New Cryst. Struct.* **1972**, *136*, 273–281. [CrossRef]
24. Baur, W.H.; Khan, A.A. Rutile-Type Compounds. IV. SiO_2, GeO_2 and a Comparison with Other Rutile-Type Structures. *Acta Crystallogr. Sect. B* **1971**, *27*, 2133–2139. [CrossRef]
25. Abbott, E.J.; Firestone, F.A. Specifying Surface Quality. *Mech. Eng.* **1933**, *55*, 569–572.
26. Whitehouse, D.J. *Handbook of Surface and Nanometrology*, 2nd ed.; CRC Press: New York, NY, USA, 2002; ISBN 9781420034196.
27. Solaymani, S.; Yadav, R.P.; Ţălu, Ş.; Achour, A.; Rezaee, S.; Nezafat, N.B. Averaged power spectrum density, fractal and multifractal spectra of Au nanoparticles deposited onto annealed TiO_2 thin films. Opt. *Quantum Electron.* **2020**, *52*, 491. [CrossRef]
28. Arman, A.; Ţălu, Ş.; Luna, C.; Ahmadpourian, A.; Naseri, M.; Molamohammadi, M. Micromorphology Characterization of Copper Thin Films by AFM and Fractal Analysis. *J. Mater. Sci. Mater. Electron.* **2015**, *26*, 9630–9639. [CrossRef]
29. Korpi, A.R.G.; Ţălu, Ş.; Bramowicz, M.; Arman, A.; Kulesza, S.; Pszczolkowski, B.; Jurečka, S.; Mardani, M.; Luna, C.; Balashabadi, P.; et al. Minkowski functional characterization and fractal analysis of surfaces of titanium nitride films. *Mater. Res. Express* **2019**, *6*, 086463. [CrossRef]
30. Grossi, M.; Riccò, B. Electrical Impedance Spectroscopy (EIS) for Biological Analysis and Food Characterization: A Review. *J. Sensors Sens. Syst.* **2017**, *6*, 303–325. [CrossRef]

Review

Application of Laser Treatment in MOS-TFT Active Layer Prepared by Solution Method

Nanhong Chen [1], Honglong Ning [1], Zhihao Liang [1], Xianzhe Liu [2], Xiaofeng Wang [3], Rihui Yao [1,*], Jinyao Zhong [1], Xiao Fu [1], Tian Qiu [4,*] and Junbiao Peng [1]

1 Institute of Polymer Optoelectronic Materials and Devices, State Key Laboratory of Luminescent Materials and Devices, South China University of Technology, Guangzhou 510640, China; chen-nanhong@foxmail.com (N.C.); ninghl@scut.edu.cn (H.N.); 201530291443@mail.scut.edu.cn (Z.L.); 202010103138@mail.scut.edu.cn (J.Z.); 201630343721@mail.scut.edu.cn (X.F.); psjbpeng@scut.edu.cn (J.P.)
2 Research Center of Flexible Sensing Materials and Devices, School of Applied Physics and Materials, Wuyi University, Jiangmen 529020, China; msliuxianzhe@mail.scut.edu.cn
3 Institute of Semiconductors, Chinese Academy of Sciences, Beijing 100083, China; wangxiaofeng@semi.ac.cn
4 Department of Intelligent Manufacturing, Wuyi University, Jiangmen 529020, China
* Correspondence: yaorihui@scut.edu.cn (R.Y.); qiutian@ustc.edu (T.Q.)

Abstract: The active layer of metal oxide semiconductor thin film transistor (MOS-TFT) prepared by solution method, with the advantages of being a low cost and simple preparation process, usually needs heat treatment to improve its performance. Laser treatment has the advantages of high energy, fast speed, less damage to the substrate and controllable treatment area, which is more suitable for flexible and large-scale roll-to-roll preparation than thermal treatment. This paper mainly introduces the basic principle of active layer thin films prepared by laser treatment solution, including laser photochemical cracking of metastable bonds, laser thermal effect, photoactivation effect and laser sintering of nanoparticles. In addition, the application of laser treatment in the regulation of MOS-TFT performance is also described, including the effects of laser energy density, treatment atmosphere, laser wavelength and other factors on the performance of active layer thin films and MOS-TFT devices. Finally, the problems and future development trends of laser treatment technology in the application of metal oxide semiconductor thin films prepared by solution method and MOS-TFT are summarized.

Keywords: solution method; laser treatment; active layer; metal oxide semiconductor thin film transistor

Citation: Chen, N.; Ning, H.; Liang, Z.; Liu, X.; Wang, X.; Yao, R.; Zhong, J.; Fu, X.; Qiu, T.; Peng, J. Application of Laser Treatment in MOS-TFT Active Layer Prepared by Solution Method. *Micromachines* **2021**, *12*, 1496. https://doi.org/10.3390/mi12121496

Academic Editor: Chengyuan Dong

Received: 10 November 2021
Accepted: 27 November 2021
Published: 30 November 2021

1. Introduction

At present, new display technology products are endlessly emerging. People continue to have higher requirements for the characteristics of display devices, such as high resolution, thin, flexible, transparent, rich color and so on. Metal oxide semiconductor thin film transistor (MOS-TFT) has the advantages of high mobility (1–100 cm^2/Vs) and good film uniformity [1–5]. It has become a strong competitor in the display backplane industry represented by active matrix liquid crystal display and active matrix organic light emitting diode.

Thin film transistor is a kind of field effect transistor. TFT devices are usually composed of active layers, insulating layers, gate electrodes, source electrodes and drain electrodes, the common TFT device structure is shown in Figure 1. In TFT, the material that plays the most important role is the semiconductor active layer. According to the difference of semiconductor active layer materials, TFT can be divided into the following four categories: a-Si TFT, p-Si TFT, OTFT and MOS-TFT [6–9]. Among them, MOS-TFT has the advantages of high field effect mobility, high uniformity, good electrical stability and high transparency, which is suitable for the future display preparation requirements, such as large size and flexibility [1,4,10].

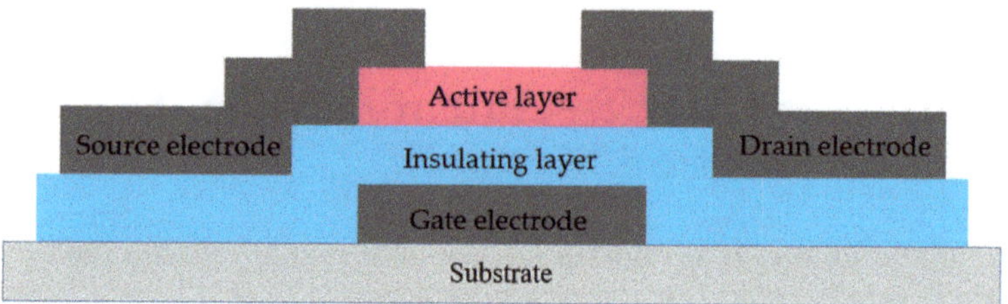

Figure 1. Schematic diagram of TFT device structure.

Solution-processed deposition offers the advantages of a simple process, high-throughput, high material utilization rate, and easy control of chemical components, which provides the possibility for large-area preparation of metal oxide semiconductor [10–14]. In the study of solution preparation of MOS-TFT, the active layer is mainly made of precursor prepared by sol–gel [15,16] or nanoparticles (NPs) dispersed in carrier solvent [17–19], which are deposited on the substrate by spin coating method, inkjet printing method and so on. Whether the thin films are prepared by sol–gel method or nano-particle method, the precursors or nanostructures usually need postprocessing to improve their properties [20]. The typical process of preparing metal oxide semiconductor thin films and corresponding TFT devices by solution method is shown in Figure 2.

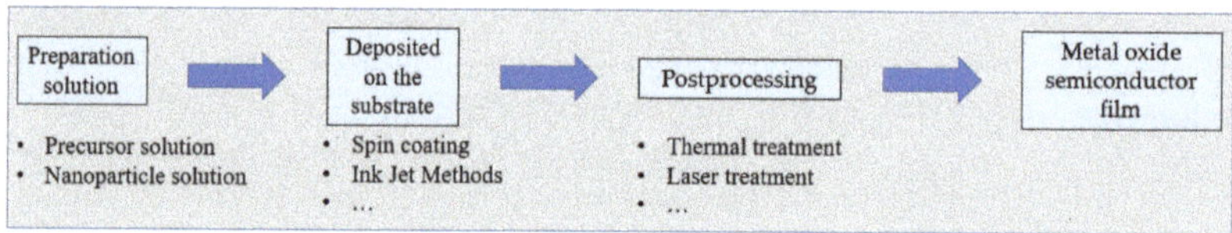

Figure 2. Schematic diagram indicating a typical solution process synthesis of metal oxide semiconductor thin films and the corresponding TFT devices.

The traditional thermal treatment process has some disadvantages, such as high energy consumption, long treatment time, high process temperature and incompatibility of flexible substrates [21]. In order to solve this problem, various studies have tried to reduce the treatment temperature by compensating for other energy sources (for example, optical, chemical and physical methods), rather than reducing the activation energy. Many researchers reduce the treatment temperature of the active layer and MOS-TFT prepared by solution method by microwave treatment [22–24], plasma treatment [23,25], ozone ultraviolet treatment (UV) [26–30], high pressure treatment [22,23,31,32], water based/hydrolysis [28,33], low temperature steam treatment [34] and so on. The common process parameters of low temperature treatment are shown in Table 1. However, these methods are not suitable for large-scale roll-to-roll (R2R) processes. On the R2R production line, the treatment time is limited by the length of the on-line curing furnace. For example, for the speed of 1 m min^{-1} and the oven length of 5 m, the curing time of each treatment layer is limited to 5 min [27,35–37], while these methods require a longer treatment time.

As a new treatment technology in the field of flexible, printing and wearable devices, laser treatment effectively avoids the shortcomings of other treatment methods, such as high energy consumption, long processing time, high process temperature, incompatibility with flexible substrate, only the whole device being treatable without the active layer being treated accurately. Laser treatment can effectively treat precursor films or nanoparticles through high-energy radiation and absorption of high-energy photons. By adjusting the laser processing parameters, such as laser intensity, pulse width and scanning speed, the energy input into the film can be accurately controlled to achieve the desired thermal effect [38–43]. In addition, the heating and cooling rate of laser treatment ($>10^6$ °C s^{-1}) is several orders of magnitude higher than that of conventional heat treatment and rapid thermal treatment, so that the thin films can be processed quickly with minimal energy loss [44]. Laser treatment is a top-down treatment technology and the treatment position can be accurately controlled, so the treatment area can be limited to a specific range of in-plane and thickness direction, and the thin films and nanostructures can be selectively treated to improve the properties of thin films and MOS-TFT without affecting the substrate and adjacent materials [43,45–47]. Common laser treatment equipment is shown in Figure 3 [48].

Table 1. Low temperature treatment process parameters.

Treatment Method	Treatment Temperature	Treatment Time	Refs.
Microwave treatment	>180 °C	>30 min	[22–24]
Plasma treatment	>300 °C	>20 min	[23,25]
Ozone ultraviolet treatment	>120 °C	>5 min	[26–30]
High pressure treatment	>220 °C	>1 h	[22,23,31,32]
Water based/hydrolysis	>230 °C	>2 h	[28,33]
Low temperature steam treatment	>220 °C	>1 h	[34]
Laser treatment	>95 °C	<5 min	[38–43]

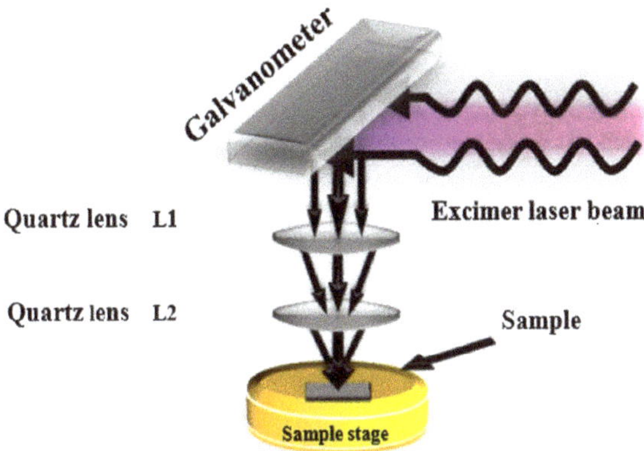

Figure 3. Schematic diagram of laser treatment device [48].

Laser treatment technology has many advantages and has made remarkable achievements in the application of active layer thin films and MOS-TFT devices prepared by solution method. However, there remains some shortcomings in the related research, such as less application on flexible substrates, less research on the influence of laser frequency and pulse number, and so on, which need to be further studied and improved.

2. Mechanism of Laser Treatment

Laser is a kind of high-energy beam with monochromaticity, coherence and collimation produced by stimulated emission process. In the process of laser treatment, the thin films are treated effectively through the thermal effect and photochemical reaction caused by the absorption of high-energy photons.

Lasers can be divided into solid-state lasers and excimer lasers according to working substances. The output beam of solid-state laser is usually Gaussian beam, and its energy curve is similar to Gaussian function curve. In contrast to solid-state lasers, the output beam of excimer lasers is usually flat-topped beam, and its energy density distribution is almost the same in a certain region.

According to the pulse width, laser can usually be divided into nanosecond laser, femtosecond laser and picosecond laser. Compared with nanosecond laser, femtosecond laser and picosecond laser can provide ultrashort pulse and low-energy high transient intensity, avoid damage to surrounding materials, and minimize thermal diffusion zone and light diffraction in ablated materials in high-resolution pattern making [49–51].

The laser wavelength and the band gap width of the material together determine the laser absorption mechanism of the material. The shorter the laser wavelength, the higher the laser photon energy. When the laser photon energy is higher than the material band gap, single photon absorption is the main mechanism of exciting valence electrons to the conduction band [51]. When the photon energy is lower than the material band gap or the single-photon absorption is suppressed by band filling, it is mainly multiphoton absorption [51].

2.1. Active Layer Thin Films Prepared by Sol-Gel Method

Unlike metal oxide films and MOS-TFT prepared by vacuum method (such as magnetron sputtering), there are impurities such as dissolved metal ligands (e.g., alkoxides, nitrates, chlorides), condensation by-products (e.g., water, alcohol), solvents and stabilizers in the precursor films prepared by solution method [29]. These impurities act as traps and hinder the effective formation of metal oxide framework, which plays the role of carrier channel. Therefore, in order to prepare high quality thin films, the removal of impurities is crucial. In order to prepare high quality metal oxide thin films, it is necessary not only to remove the impurities contained in the precursors, but also to provide enough energy to promote the Polycondensation and the densification of the thin films to form the metal-oxygen-metal (M-O-M) lattice structure [52–55]. The improvement of the lattice structure and the densification of the thin film help to reduce the traps and the potential barrier, increase the carrier mobility, and then improve the device performance of MOS-TFT [43,47,53,54,56–60].

Laser treatment can effectively remove the impurities in the precursor film and promote the formation of lattice network [29,54,55,61–65]. There are usually three mechanisms for the interaction between laser and precursor film: (1) thermal effect of laser; (2) photochemical cleavage of metastable bonds; and (3) photochemical effect. The process of thermal effect of laser usually includes: (1) carrier excitation; (2) carrier–carrier scattering, carrier–phonon scattering, and energy transfer to the lattice due to spontaneous phonon emission; (3) when the carrier and lattice reach equilibrium, the film is heated, as shown in Figure 4 [51,66]. High energy photons can induce the photochemical cleavage of chemical bonds related to metal alkoxy and carbon impurities and promote the subsequent reorganization of metal oxide frames [27,55,63]. These mechanisms are also observed in other light-assisted methods. However, the difference between laser processing and other photo-assisted methods that take a longer time is that laser combines these photochemical effects with laser-induced high temperature heating to provide additional local heat energy. Therefore, the laser treatment can effectively decompose the impurities related to the precursor, remove the metal oxide defects, and reorder the metal oxide structure instantly (<100 ns) at a lower substrate temperature (RT). Laser acting on thin films can not only produce thermal effect through instantaneous high energy radiation, but also produce photoactivation effect by high energy photons [43,53,67,68]. The photoactivation effect is that the residual metal ligands in the precursor films are photolyzed by high energy photons to produce free radicals. The free radicals mediate the reaction to form the M-O-M lattice structure and decompose the chemical impurities into small gas molecules [29], as shown in Figure 5.

Juan et al. used KrF excimer laser with wavelength of 248 nm to treat IZO-TFT and found that excimer laser treatment effectively removed carbon impurities related to the precursor and improved the lattice structure [43]. Chen et al. used a femtosecond laser with a wavelength of 800 nm to treat IZO-TFT. The high-energy photons generated by the laser induced the photo assisted condensation reaction, resulting in the formation of metal oxide bonds by metal hydroxides, and dehydroxylation reaction at the same time to remove residual impurities [62]. Dellis et al. treated In_2O_3 thin films with KrF excimer laser and characterized them by X-ray photoelectron spectroscopy (XPS). The degree of conversion of initial precursors to metal oxides was evaluated by the ratio of In-O to In-OH bonds. After laser treatment, the proportion of In-O bond increased significantly, while the proportion of In-OH bond decreased. The results show that laser treatment can effectively promote the transformation of precursors to metal oxides [69]. Fei et al. used femtosecond laser treatment to treat IZO-TFT. They proposed that laser treatment can break In-O and Zn-O bonds and form metal oxygen lattice structures such as In-O-Zn-O or Zn-O-In-O under thermal effect [52].

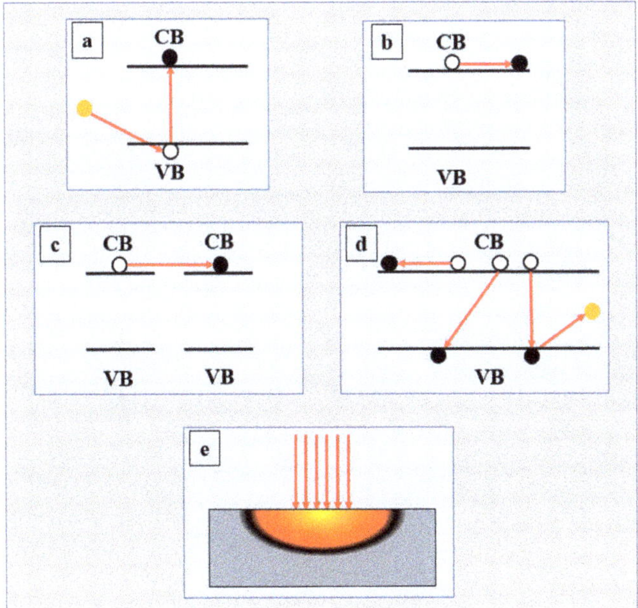

Figure 4. Schematic diagram of laser thermal effect: (**a**) photon absorption and carrier excitation; (**b**) carrier–carrier scattering; (**c**) carrier-phonon scattering; (**d**) carrier recombination; (**e**) thermal effect and thermal diffusion.

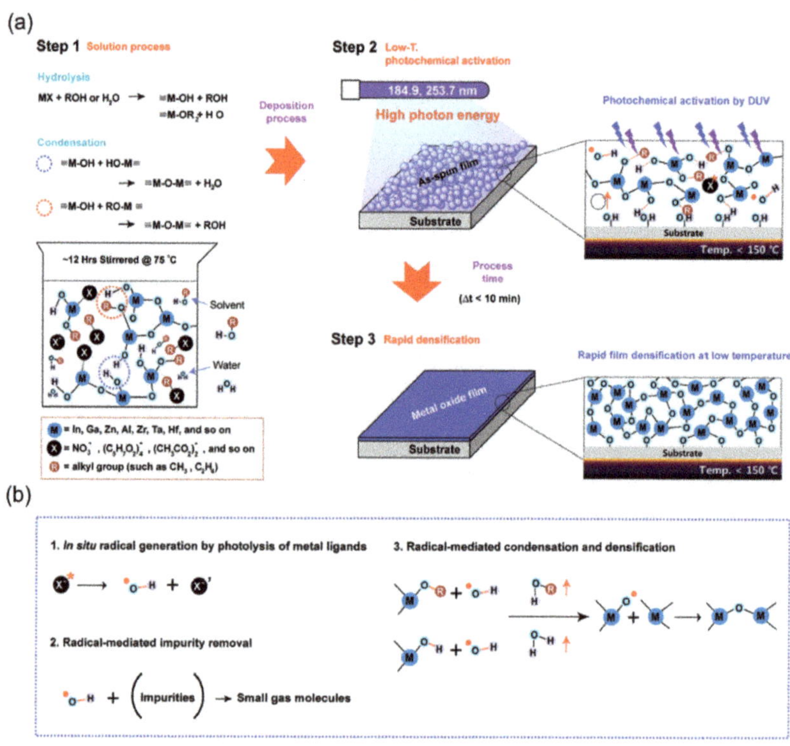

Figure 5. Photoactivation of sol–gel metal oxide materials and the proposed mechanism: (**a**) overall schematic illustration of the rapid low-temperature photoactivation of various sol–gel metal oxide films; (**b**) proposed physicochemical mechanism of the rapid low-temperature photoactivation process via photochemical activation (direct photodecomposition of impurities, in situ radical formation, enhancement of rapid condensation and densification) [29].

2.2. Active Layer Thin Films Prepared by Nano-Particle Method

In addition to sol–gel method, nano-particle method is another common solution method to prepare MOS-TFT active layer. Nanoparticles are prepared by coprecipitation or hydrothermal method and deposited by inkjet printing or rotary coating [44,70]. Laser treatment can provide high temperature up to the melting point of nanoparticles, sinter nanoparticles and form semiconductor films [71,72]. Qion et al. prepared AZO thin films by rotary coating method, and studied the laser sintering process of nano-particles. According to their simulation study, they proposed that in the process of interaction between laser and nanoparticles, the contact zone between nanoparticles is first heated to form hot spots, and then the heat spreads to the interior of the particles and adjacent particles. The hot spots in the contact zone promote the surface melting and merging of the nanoparticles, increase the grain size, change the grain shape and compress the internal gap, and finally form a continuous dense film [71]. Lee et al. prepared ZnO-TFT by nano-particle method and treated with yttrium vanadate (Nd:YVO_4) picosecond (ps) ultraviolet laser. Their study found that before laser treatment, particles and nano-pores were observed, and the thickness of the film was 175 nm. After laser treatment, the grains are melted, the voids are reduced, and the thickness of the film is reduced to 95 nm, indicating that the film is densified by laser treatment, as shown in Figure 6 [73].

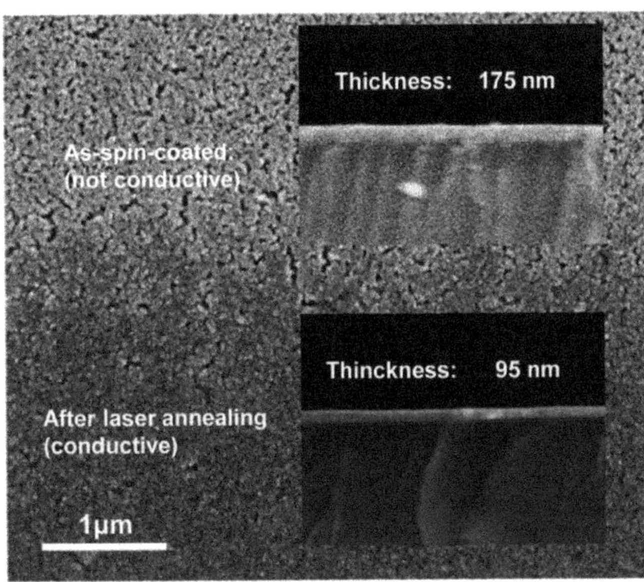

Figure 6. SEM images of ZnO films before and after laser treatment [73].

3. Application of Laser Treatment in MOS-TFT Performance Control

Combined with the properties of the thin film (film thickness, composition, absorption spectrum, etc.), the physical, optical, electrical and chemical properties of the thin film and MOS-TFT can be adjusted by changing the laser processing parameters (energy density, frequency, treatment atmosphere, etc.). Table 2 summarizes the examples of laser processing of metal oxide semiconductor thin film transistors.

Table 2. Examples of laser treatment of MOS-TFT.

Channel Material	Solution Type	Laser Wavelength (nm)	μ (cm^2 V^{-1} s^{-1})	SS (V dec^{-1})	On/Off Ratio	Ref.
IGZO	Sol-gel	355	7.65			[74]
IGZO	Sol-gel	800	4.24	0.91	7.2×10^5	[75]
ZnO	NPs	355	0.5		1.7×10^6	[53]
IGZO	NPs	355	7.65		2.71×10^6	[53]
IGZO	Sol-gel	1064	1.5		1.29×10^6	[76]
In_2O_3	Sol-gel	700	10.03 ± 0.64	1.44 ± 0.37	3.4×10^5	[63]
In_2O_3	Sol-gel	248	13		10^6	[69]
IZO	Sol-gel	800	3.75	1.21	1.77×10^5	[52]
IZO	Sol-gel	248	0.58			[47]
ZnO	NPs	355	3.01	1.8	10^5	[73]

3.1. Laser Energy Density

Laser energy density is an important factor affecting the lattice structure of metal oxide films. When the laser energy density is too low, the film may not be treated effectively, and when the energy density is too high, it may also have an adverse impact on the performance of the device [77]. Chen et al. prepared IGZO-TFT and treated it with femtosecond laser with wavelength of 800 nm and energy density of 20, 35, 80, 112 and 130 mJ/cm^2. Their research found that the films treated at 20 mJ/cm^2 energy density will produce serious defects and trap states due to the incomplete transformation of precursors to metal oxide lattice, and the devices do not have TFT characteristics. With the increase of laser energy density, the performance of TFT devices is improved, and the best device performance is obtained at 112 mJ/cm^2. When the laser energy density increases to 130 mJ/cm^2, the performance of the device decreases, as shown in Figure 7 [75].

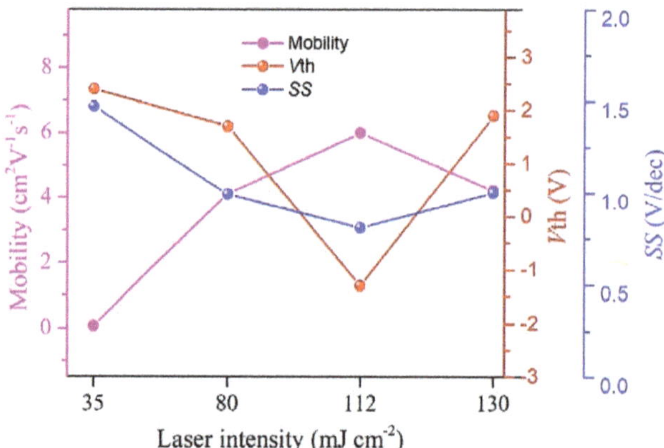

Figure 7. Variation in device performance results of IGZO-TFTs as a function of laser intensity [75].

There is usually a certain energy threshold in metal oxide thin films, and the laser energy exceeding the threshold will induce the recrystallization or grain growth of the thin films [68,78,79]. The grain size usually increases with the increase of laser energy density [80,81]. For polycrystalline thin films, the bottleneck of field effect mobility usually occurs at grain boundaries, so reducing the number of grain boundaries and increasing grain size by laser treatment can improve the device performance [60,82–85]. Nagase et al. studied the effects of laser energy density and film thickness on the properties of ZnO films. Their research found that two kinds of crystal ZnO films were obtained under different laser energy density and different film thickness. Low energy density produces low crystallinity with weak orientation, while high energy density produces high crystallinity with strong orientation, and the threshold of energy density increases with the increase of film thickness [86]. Yang et al. prepared ZnO-TFT and treated it with a Nd:YAG laser with a wavelength of 355 nm. It is found that laser treatment can improve the crystallinity of ZnO materials, and the mobility of TFT devices is increased by more than 2.5 times (0.19 to 0.49 cm^2/Vs) as shown in Figure 8 [58].

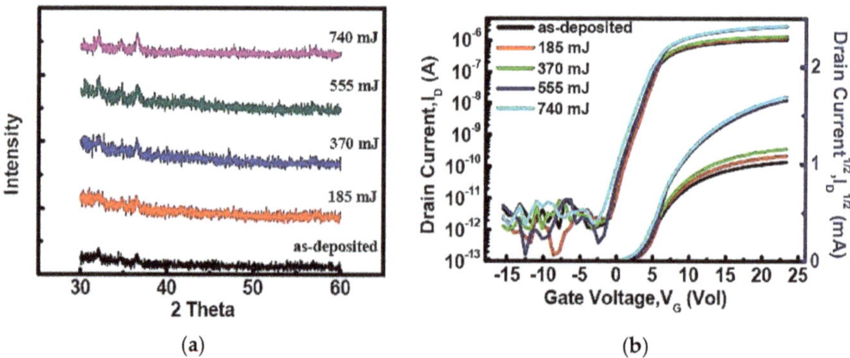

Figure 8. Crystallization degree of ZnO thin films and transmission characteristics of ZnO-TFT devices under different laser energy densities: (**a**) degree of crystallization; (**b**) transmission characteristics [58].

3.2. Treatment Atmosphere

Laser treatment in air has the advantages of low cost, simple process and more suitable for large-area manufacturing, but the moisture and oxygen in air will affect the properties of the film [40,87]. During laser treatment, the film surface will be heated to a temperature sufficient to destroy the M-O bond and form an oxygen vacancy [88–90]. If treated in an air atmosphere, the oxygen in the air will oxidize the metal elements in the film again to form an M-O bond. The destruction rate of M-O bond and the oxidation rate of metal elements together determine the concentration of oxygen vacancies in the thin films. Usually, the formation of oxygen vacancies is often accompanied by the generation of electrons, which increases the carrier concentration of metal oxide films [68,91]. The increase of oxygen vacancy concentration causes high carrier concentration to form an electron transport path near the conduction band, thus increasing the mobility of TFT devices and reducing the threshold voltage [76,92–95]. However, too high oxygen vacancy concentration may lead to high leakage current of TFT devices due to high carrier concentration, which reduces the device performance [52]. Therefore, by adjusting the gas atmosphere during laser treatment, the concentration of oxygen vacancies in the thin films can be effectively controlled and the electrical properties of MOS-TFT devices can be improved. Lee et al. studied the effects of ambient atmosphere and argon atmosphere on the laser-treated ZnO thin films. Their study found that the oxygen vacancy content of the films treated in argon atmosphere was significantly higher than that in the ambient atmosphere [73]. Juan et al. treated IZO-TFT with KrF excimer laser in air atmosphere and vacuum. Their study found that laser treatment in vacuum can inhibit the absorption of extra water and excess oxygen from the atmosphere, and the device mobility is higher than that of laser treatment in air, as shown in Figure 9 [43].

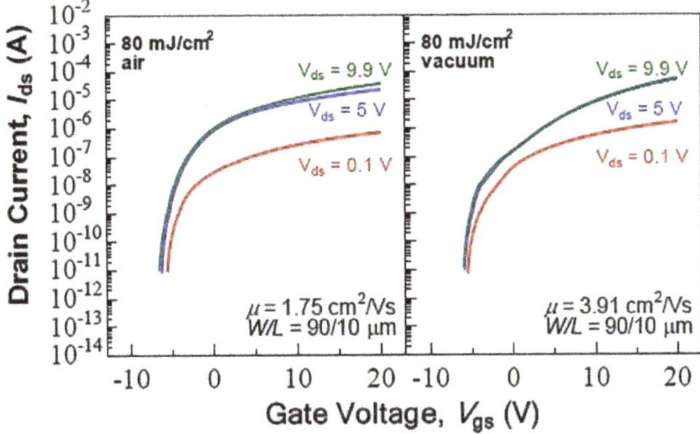

Figure 9. Performance of IZO-TFT devices treated by laser in different atmospheres [43].

3.3. Laser Wavelength

Due to laser processing is a top-down process, according to Beer-Lambert law, the radiation intensity of laser attenuates in the film [96,97]. The shorter the laser wavelength is, the shallower the penetration depth is. Therefore, in the process of laser treatment, temperature gradients are easy to exist in the film, resulting in differences in the properties of regions with different depths of the film. This effect is particularly significant in the films treated by ultraviolet wavelength lasers [55,98,99]. Kwon et al. prepared ZTO thin films by sol–gel method and treated them with KrF excimer laser. They performed high-resolution chemical and microstructure analysis of the films. Their study found that during the UV laser treatment, the top temperature of the ZTO film is much higher than that in the deep region. This temperature gradient makes the Zn element enriched in the surface region and the Sn element enriched in the bottom region [55]. Sandu et al. found that due to the different penetration thickness of SnO_2 thin films by 193 nm and 248 nm laser (66 nm and 148 nm, respectively), the crystallization effect is different, and the crystal gradient of the thin film irradiated by 193 nm laser is more obvious [80,81]. In contrast to the UV laser, the infrared laser has a long wavelength and a large penetration depth in the film, which can heat the film more evenly. However, due to its deep penetration depth, when applied to flexible MOS-TFT, the flexible substrate may be damaged by a large number of high-energy photons, which will affect the performance of MOS-TFT [9,51,75,100]. Chen et al. fabricated MOS-TFT devices and embedded dielectric mirrors (DMs) in them. Their research shows that the DMs can effectively prevent the penetration of high energy photons into the PEN substrate, thus avoiding the damage to the substrate and significantly improving the performance of the device, as shown in Figure 10 [75].

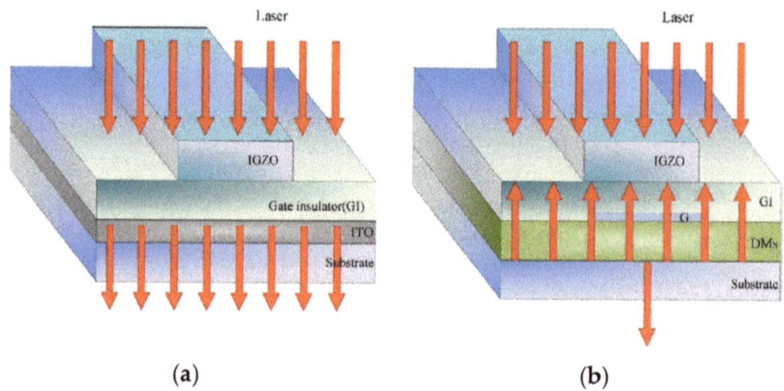

Figure 10. Protective effect of DMs on flexible substrate: (**a**) without DMs; (**b**) with DMs [75].

4. Conclusions and Prospects

The excellent compatibility between laser processing technology and solution method has been widely recognized. In the sol–gel method, laser treatment can effectively remove impurities in the precursor films and promote the formation of lattice networks. In the nano-particle method, the nano-particles are sintered effectively by laser treatment to form a continuous and dense film. The performance parameters of thin film and MOS-TFT can be effectively improved by adjusting various parameters in the process of laser processing. Although laser processing technology has made remarkable achievements in the application of active layer thin films and MOS-TFT devices prepared by solution method, there are still some deficiencies in the production of flexible, large-size, low-cost MOS-TFT and further improving the performance of MOS-TFT devices, which need to be further studied and improved. These include (1) less research on flexible devices; (2) less research combined with other low temperature treatment processes. (3) present research only focused on the effect of different laser energy density on thin films; there are few studies on other parameters of laser treatment, such as pulse number, frequency and so on; (4) the research on the mechanism and physical model of the interaction between laser and thin film is not deep enough. In the future, by using different flexible substrates, adjusting the process parameters of laser processing, and combining laser processing with other low temperature treatment processes, study of the application of laser processing technology in active layer thin films and MOS-TFT devices prepared by solution method, so as to promote the development of flexible large size display technology should continue.

Author Contributions: Funding acquisition, H.N. and R.Y.; investigation, N.C.; project administration, Z.L. and J.P.; resources, X.F., X.L., H.N., T.Q., R.Y. and X.W.; supervision, H.N.; validation, J.Z.; writing–original draft, N.C.; writing–review & editing, H.N., T.Q. and R.Y. All authors have read and agreed to the published version of the manuscript.

Funding: This research was funded by Key-Area Research and Development Program of Guangdong Province (No.2020B010183002), National Natural Science Foundation of China (Grant No.62074059 and 22090024), Guangdong Major Project of Basic and Applied Basic Research (No.2019B030302007), Fundamental Research Funds for the Central Universities (No.2020ZYGXZR060), Ji Hua Laboratory scientific research project (X190221TF191), South China University of Technology 100 Step Ladder Climbing Plan Research Project (No.j2tw202102000) and 2021 Guangdong University Student Science and Technology Innovation Special Fund ("Climbing Plan" Special Fund) (No.pdjh2021b0036).

Conflicts of Interest: The authors declare no conflict of interest.

References

1. Yao, R.H.; Fu, X.; Li, X.Q.; Qiu, T.; Ning, H.L.; Yang, Y.X.; Lu, X.B.; Cao, X.H.; Chen, Y.; Peng, J.B. Performances of thin film transistors with Ga-doped ZnO source and drain electrodes. *J. Phys. D-Appl. Phys.* **2021**, *54*, 365101–365107. [CrossRef]
2. Hu, S.; Lu, K.; Ning, H.; Yao, R.; Gong, Y.; Pan, Z.; Guo, C.; Wang, J.; Pang, C.; Gong, Z.; et al. Study of the Correlation between the Amorphous Indium-Gallium-Zinc Oxide Film Quality and the Thin-Film Transistor Performance. *Nanomaterials* **2021**, *11*, 522. [CrossRef]
3. Liu, X.; Shiah, Y.-S.; Guo, D.; Ning, H.; Zhang, X.; Chen, J.; Fu, X.; Wang, Y.; Yao, R.; Peng, J. Origin of bias-stress and illumination instability in low-cost, wide-bandgap amorphous Si-doped tin oxide-based thin-film transistors. *J. Phys. D-Appl. Phys.* **2020**, *53*, 235102–235107. [CrossRef]
4. Lu, K.; Yao, R.; Xu, W.; Ning, H.; Zhang, X.; Zhang, G.; Li, Y.; Zhong, J.; Yang, Y.; Peng, J. Alloy-Electrode-Assisted High-Performance Enhancement-Type Neodymium-Doped Indium-Zinc-Oxide Thin-Film Transistors on Polyimide Flexible Substrate. *Research* **2021**, *2021*. [CrossRef]
5. Lu, K.; Zhang, J.; Guo, D.; Xiang, J.; Lin, Z.; Zhang, X.; Wang, T.; Ning, H.; Yao, R.; Peng, J. High-Performance and Flexible Neodymium-Doped Indium-Zinc-Oxide Thin-Film Transistor with All Copper Alloy Electrodes. *IEEE Electron. Device Lett.* **2020**, *41*, 417–420. [CrossRef]
6. Sameshima, T.; Usui, S.; Sekiya, M. XeCl excimer laser annealing used in the fabrication of poly-Si TFTs. *IEEE Electron. Device Lett.* **1986**, *1986*, 276–278. [CrossRef]
7. Kang, D.H.; Park, M.K.; Jang, J.; Chang, Y.J.; Oh, J.H.; Choi, J.B.; Kim, C.W. Active-matrix organic light-emitting diode using inverse-staggered poly-Si TFTs with a center-offset gated structure. *J. Soc. Inf. Disp.* **2010**, *18*, 122–127. [CrossRef]
8. Kugimiya, T.; Yoneda, Y.; Kusumoto, E.; Gotoh, H.; Ochi, M.; Kawakami, N.; Sid. Single layer Al-Ni-La-Si interconnections for source and drain of LTPS-TFT LCDs using direct contacts with both ITO and poly-Si. In Proceedings of the International Symposium of the Society-for-Information-Display (SID 2008), Los Angeles, CA, USA, 18–23 May 2008.
9. Nomura, K.; Ohta, H.; Takagi, A.; Kamiya, T.; Hirano, M.; Hosono, H. Room-temperature fabrication of transparent flexible thin-film transistors using amorphous oxide semiconductors. *Nature* **2004**, *432*, 488–492. [CrossRef]
10. Ye, Q.; Zhang, X.; Guo, D.; Xu, W.; Ning, H.; Qiu, T.; Li, J.; Hou, D.; Yao, R.; Peng, J. Preparation of Highly Transparent (at 450–800 nm) SnO_2 Homojunction by Solution Method and Its Photoresponse. *Coatings* **2020**, *10*, 399. [CrossRef]
11. Cai, W.; Zhu, Z.; Wei, J.; Fang, Z.; Ning, H.; Zheng, Z.; Zhou, S.; Yao, R.; Peng, J.; Lu, X. A Simple Method for High-Performance, Solution-Processed, Amorphous ZrO_2 Gate Insulator TFT with a High Concentration Precursor. *Materials* **2017**, *10*, 972. [CrossRef]
12. Liang, Z.; Zhou, S.; Cai, W.; Fu, X.; Ning, H.; Chen, J.; Yuan, W.; Zhu, Z.; Yao, R.; Peng, J. Zirconium-Aluminum-Oxide Dielectric Layer with High Dielectric and Relatively Low Leakage Prepared by Spin-Coating and the Application in Thin-Film Transistor. *Coatings* **2020**, *10*, 282. [CrossRef]
13. Ning, H.; Zhang, X.; Wang, S.; Yao, R.; Liu, X.; Hou, D.; Ye, Q.; Li, J.; Huang, J.; Cao, X.; et al. Preparation and optimization of SnOx thin film by solution method at low temperature. *Superlattices Microstruct.* **2020**, *139*, 106400–106408. [CrossRef]
14. Zhu, Z.; Zhang, J.; Guo, D.; Ning, H.; Zhou, S.; Liang, Z.; Yao, R.; Wang, Y.; Lu, X.; Peng, J. Functional Metal Oxide Ink Systems for Drop-on-Demand Printed Thin-Film Transistors. *Langmuir* **2020**, *36*, 8655–8667. [CrossRef]
15. Natsume, Y.; Sakata, H. Zinc oxide films prepared by sol-gel spin-coating. *Thin Solid Films* **2000**, *372*, 30–36. [CrossRef]
16. Ohyama, M.; Kozuka, H.; Yoko, T. Sol-gel preparation of ZnO films with extremely preferred orientation along (002) plane from zinc acetate solution. *Thin Solid Films* **1997**, *306*, 78–85. [CrossRef]
17. Sun, B.; Sirringhaus, H. Solution-processed zinc oxide field-effect transistors based on self-assembly of colloidal nanorods. *Nano Lett.* **2005**, *5*, 2408–2413. [CrossRef]
18. Berber, M.; Bulto, V.; Kliss, R.; Hahn, H. Transparent nanocrystalline ZnO films prepared by spin coating. *Scr. Mater.* **2005**, *53*, 547–551. [CrossRef]
19. Ismail, B.; Abaab, M.; Rezig, B. Structural and electrical properties of ZnO films prepared by screen printing technique. *Thin Solid Films* **2001**, *383*, 92–94. [CrossRef]
20. Subramanian, V.; Bakhishev, T.; Redinger, D.; Volkman, S.K. Solution-Processed Zinc Oxide Transistors for Low-Cost Electronics Applications. *J. Disp. Technol.* **2009**, *5*, 525–530. [CrossRef]
21. Banger, K.K.; Peterson, R.L.; Mori, K.; Yamashita, Y.; Leedham, T.; Sirringhaus, H. High Performance, Low Temperature Solution-Processed Barium and Strontium Doped Oxide Thin Film Transistors. *Chem. Mat.* **2014**, *26*, 1195–1203. [CrossRef]

22. Ahn, B.D.; Jeon, H.J.; Sheng, J.Z.; Park, J.; Park, J.S. A review on the recent developments of solution processes for oxide thin film transistors. *Semicond. Sci. Technol.* **2015**, *30*, 15. [CrossRef]
23. Xu, W.Y.; Li, H.; Xu, J.B.; Wang, L. Recent Advances of Solution-Processed Metal Oxide Thin-Film Transistors. *ACS Appl. Mater. Interfaces* **2018**, *10*, 25878–25901. [CrossRef] [PubMed]
24. Jun, T.; Song, K.; Jeong, Y.; Woo, K.; Kim, D.; Bae, C.; Moon, J. High-performance low-temperature solution-processable ZnO thin film transistors by microwave-assisted annealing. *J. Mater. Chem.* **2011**, *21*, 1102–1108. [CrossRef]
25. Nayak, P.K.; Hedhili, M.N.; Cha, D.K.; Alshareef, H.N. High performance solution-deposited amorphous indium gallium zinc oxide thin film transistors by oxygen plasma treatment. *Appl. Phys. Lett.* **2012**, *100*, 4. [CrossRef]
26. Su, B.Y.; Chu, S.Y.; Juang, Y.D.; Chen, H.C. High-performance low-temperature solution-processed InGaZnO thin-film transistors via ultraviolet-ozone photo-annealing. *Appl. Phys. Lett.* **2013**, *102*, 4. [CrossRef]
27. Kim, Y.H.; Heo, J.S.; Kim, T.H.; Park, S.; Yoon, M.H.; Kim, J.; Oh, M.S.; Yi, G.R.; Noh, Y.Y.; Park, S.K. Flexible metal-oxide devices made by room-temperature photochemical activation of sol-gel films. *Nature* **2012**, *489*, 128–132. [CrossRef]
28. Park, S.; Kim, C.H.; Lee, W.J.; Sung, S.; Yoon, M.H. Sol-gel metal oxide dielectrics for all-solution-processed electronics. *Mater. Sci. Eng. Rep.* **2017**, *114*, 1–22. [CrossRef]
29. Park, S.; Kim, K.H.; Jo, J.W.; Sung, S.; Kim, K.T.; Lee, W.J.; Kim, J.; Kim, H.J.; Yi, G.R.; Kim, Y.H.; et al. In-Depth Studies on Rapid Photochemical Activation of Various Sol-Gel Metal Oxide Films for Flexible Transparent Electronics. *Adv. Funct. Mater.* **2015**, *25*, 2807–2815. [CrossRef]
30. Carlos, E.; Branquinho, R.; Kiazadeh, A.; Barquinha, P.; Martins, R.; Fortunato, E. UV-Mediated Photochemical Treatment for Low-Temperature Oxide-Based Thin-Film Transistors. *ACS Appl. Mater. Interfaces* **2016**, *8*, 31100–31108. [CrossRef]
31. Kim, S.J.; Yoon, S.; Kim, H.J. Review of solution-processed oxide thin-film transistors. *Jpn. J. Appl. Phys.* **2014**, *53*, 10. [CrossRef]
32. Rim, Y.S.; Jeong, W.H.; Kim, D.L.; Lim, H.S.; Kim, K.M.; Kim, H.J. Simultaneous modification of pyrolysis and densification for low-temperature solution-processed flexible oxide thin-film transistors. *J. Mater. Chem.* **2012**, *22*, 12491–12497. [CrossRef]
33. Banger, K.K.; Yamashita, Y.; Mori, K.; Peterson, R.L.; Leedham, T.; Rickard, J.; Sirringhaus, H. Low-temperature, high-performance solution-processed metal oxide thin-film transistors formed by a 'sol-gel on chip' process. *Nat. Mater.* **2011**, *10*, 45–50. [CrossRef] [PubMed]
34. Woods, K.N.; Plassmeyer, P.N.; Park, D.H.; Enman, L.J.; Grealish, A.K.; Kirk, B.L.; Boettcher, S.W.; Keszler, D.A.; Page, C.J. Low-Temperature Steam Annealing of Metal Oxide Thin Films from Aqueous Precursors: Enhanced Counterion Removal, Resistance to Water Absorption, and Dielectric Constant. *Chem. Mat.* **2017**, *29*, 8531–8538. [CrossRef]
35. Chang, J.S.; Facchetti, A.F.; Reuss, R. A Circuits and Systems Perspective of Organic/Printed Electronics: Review, Challenges, and Contemporary and Emerging Design Approaches. *IEEE J. Emerg. Sel. Top. Circuits Syst.* **2017**, *7*, 7–26. [CrossRef]
36. Leppaniemi, J.; Ojanpera, K.; Kololuoma, T.; Huttunen, O.H.; Dahl, J.; Tuominen, M.; Laukkanen, P.; Majumdar, H.; Alastalo, A. Rapid low-temperature processing of metal-oxide thin film transistors with combined far ultraviolet and thermal annealing. *Appl. Phys. Lett.* **2014**, *105*, 5. [CrossRef]
37. Carlos, E.; Dellis, S.; Kalfagiannis, N.; Koutsokeras, L.; Koutsogeorgis, D.C.; Branquinho, R.; Martins, R.; Fortunato, E. Laser induced ultrafast combustion synthesis of solution-based AlOx for thin film transistors. *J. Mater. Chem. C* **2020**, *8*, 6176–6184. [CrossRef]
38. Bharadwaja, S.S.N.; Dechakupt, T.; Trolier-McKinstry, S.; Beratan, H. Excimer laser crystallized (Pb,La)(Zr,Ti)O_3 thin films. *J. Am. Ceram. Soc.* **2008**, *91*, 1580–1585. [CrossRef]
39. Queralto, A.; del Pino, A.P.; de la Mata, M.; Tristany, M.; Obradors, X.; Puig, T.; Trolier-McKinstry, S. Ultraviolet pulsed laser crystallization of $Ba_{0.8}Sr_{0.2}TiO_3$ films on $LaNiO_3$-coated silicon substrates. *Ceram. Int.* **2016**, *42*, 4039–4047. [CrossRef]
40. Dutta Majumdar, J.; Manna, I. *Laser-Assisted Fabrication of Materials*; Springe: Berlin, Germany, 2013.
41. Kalfagiannis, N.; Siozios, A.; Bellas, D.V.; Toliopoulos, D.; Bowen, L.; Pliatsikas, N.; Cranton, W.M.; Kosmidis, C.; Koutsogeorgis, D.C.; Lidorikis, E.; et al. Selective modification of nanoparticle arrays by laser-induced self assembly (MONA-LISA): Putting control into bottom-up plasmonic nanostructuring. *Nanoscale* **2016**, *8*, 8236–8244. [CrossRef]
42. Tsakonas, C.; Cranton, W.; Li, F.; Abusabee, K.; Flewitt, A.; Koutsogeorgis, D.; Ranson, R. Intrinsic photoluminescence from low temperature deposited zinc oxide thin films as a function of laser and thermal annealing. *J. Phys. Appl. Phys.* **2013**, *46*, 9. [CrossRef]
43. Bermundo, J.P.S.; Kulchaisit, C.; Ishikawa, Y.; Fujii, M.N.; Ikenoue, H.; Uraoka, Y. Rapid photo-assisted activation and enhancement of solution-processed InZnO thin-film transistors. *J. Phys. Appl. Phys.* **2020**, *53*, 7. [CrossRef]
44. Palneedi, H.; Park, J.H.; Maurya, D.; Peddigari, M.; Hwang, G.T.; Annapureddy, V.; Kim, J.W.; Choi, J.J.; Hahn, B.D.; Priya, S.; et al. Laser Irradiation of Metal Oxide Films and Nanostructures: Applications and Advances. *Adv. Mater.* **2018**, *30*, 38. [CrossRef]
45. Hong, S.; Lee, H.; Yeo, J.; Ko, S.H. Digital selective laser methods for nanomaterials: From synthesis to processing. *Nano Today* **2016**, *11*, 547–564. [CrossRef]
46. Yu, H.; Lee, H.; Lee, J.; Shin, H.; Lee, M. Laser-assisted patterning of solution-processed oxide semiconductor thin film using a metal absorption layer. *Microelectron. Eng.* **2011**, *88*, 6–10. [CrossRef]
47. Chen, C.-N.; Huang, J.-J. Effects of excimer laser annealing on low-temperature solution based indium-zinc-oxide thin film transistor fabrication. *J. Appl. Res. Technol.* **2015**, *13*, 170–176. [CrossRef]
48. Deng, Y.; Liu, X.; Yuan, W.; Ning, H.; Tao, S.; Liu, Y.; Ou, Z.; Wang, X.; Yao, R.; Peng, J. Effect of deep UV laser treatment on silicon-doped Tin oxide thin film. *J. Soc. Inf. Disp.* **2020**, *28*, 194–203. [CrossRef]

49. Gattass, R.R.; Mazur, E. Femtosecond laser micromachining in transparent materials. *Nat. Photonics* **2008**, *2*, 219–225. [CrossRef]
50. Chichkov, B.N.; Momma, C.; Nolte, S.; von Alvensleben, F.; Tuennermann, A. Femtosecond, picosecond and nanosecond laser ablation of solids. *Appl. Phys. A* **1996**, *63*, 109–115. [CrossRef]
51. Sundaram, S.K.; Mazur, E. Inducing and probing non-thermal transitions in semiconductors using femtosecond laser pulses. *Nat. Mater.* **2002**, *1*, 217–224. [CrossRef]
52. Shan, F.; Kim, S.J. Dynamic Response Behavior of Femtosecond Laser-Annealed Indium Zinc Oxide Thin-Film Transistors. *J. Electr. Eng. Technol.* **2017**, *12*, 2353–2358. [CrossRef]
53. Yang, Y.H.; Yang, S.S.; Chou, K.S. Performance improvements of IGZO and ZnO thin-film transistors by laser-irradiation treatment. *J. Soc. Inf. Disp.* **2011**, *19*, 247–252. [CrossRef]
54. Xu, M.; Peng, C.; Yuan, Y.Y.; Li, X.F.; Zhang, J.H. Enhancing the Performance of Solution-Processed Thin-Film Transistors via Laser Scanning Annealing. *ACS Appl. Electron. Mater.* **2020**, *2*, 2970–2975. [CrossRef]
55. Kwon, J.H.; Jang, S.; Kim, H.J.; Joo, B.S.; Yu, K.N.; Choi, E.; Han, M.; Park, J.H.; Chang, Y.J. Microscopic and chemical analysis of room temperature UV laser annealing of solution-based zinc-tin-oxide thin films. *J. Anal. Sci. Technol.* **2020**, *11*, 9. [CrossRef]
56. Takagi, A.; Nomura, K.; Ohta, H.; Yanagi, H.; Kamiya, T.; Hirano, M.; Hosono, H. Carrier transport and electronic structure in amorphous oxide semiconductor, a-$InGaZnO_4$. *Thin Solid Films* **2005**, *486*, 38–41. [CrossRef]
57. Yu, X.G.; Zhou, N.J.; Smith, J.; Lin, H.; Stallings, K.; Yu, J.S.; Marks, T.J.; Facchetti, A. Synergistic Approach to High-Performance Oxide Thin Film Transistors Using a Bilayer Channel Architecture. *ACS Appl. Mater. Interfaces* **2013**, *5*, 7983–7988. [CrossRef]
58. Yang, Y.H.; Yang, S.S.; Chou, K.S. Laser-irradiated zinc oxide thin-film transistors fabricated by solution processing. *J. Soc. Inf. Disp.* **2010**, *18*, 745–748. [CrossRef]
59. Jo, G.; Ji, J.H.; Masao, K.; Ha, J.G.; Lee, S.K.; Koh, J.H. CO_2 laser annealing effects for Al-doped ZnO multilayered films. *Ceram. Int.* **2018**, *44*, S211–S215. [CrossRef]
60. Kang, J.; Jo, G.; Ji, J.H.; Koh, J.H. Improved electrical properties of laser annealed In and Ga co-doped ZnO thin films for transparent conducting oxide applications. *Ceram. Int.* **2019**, *45*, 23934–23940. [CrossRef]
61. Hwang, J.; Lee, K.; Jeong, Y.; Lee, Y.U.; Pearson, C.; Petty, M.C.; Kim, H. UV-Assisted Low Temperature Oxide Dielectric Films for TFT Applications. *Adv. Mater. Interfaces* **2014**, *1*, 9. [CrossRef]
62. Chen, C.H.; Chen, G.X.; Yang, H.H.; Zhang, G.C.; Hu, D.B.; Chen, H.P.; Guo, T.L. Solution-processed metal oxide arrays using femtosecond laser ablation and annealing for thin-film transistors. *J. Mater. Chem. C* **2017**, *5*, 9273–9280. [CrossRef]
63. Shan, F.; Sun, H.Z.; Lee, J.Y.; Pyo, S.; Kim, S.J. Improved High-Performance Solution Processed In_2O_3 Thin Film Transistor Fabricated by Femtosecond Laser Pre-Annealing Process. *IEEE Access* **2021**, *9*, 44453–44462. [CrossRef]
64. Tsay, C.Y.; Huang, T.T. Improvement of physical properties of IGZO thin films prepared by excimer laser annealing of sol-gel derived precursor films. *Mater. Chem. Phys.* **2013**, *140*, 365–372. [CrossRef]
65. Kim, H.J.; Maeng, M.J.; Park, J.H.; Kang, M.G.; Kang, C.Y.; Park, Y.; Chang, Y.J. Chemical and structural analysis of low-temperature excimer-laser annealing in indium-tin oxide sol-gel films. *Curr. Appl. Phys.* **2019**, *19*, 168–173. [CrossRef]
66. Choi, J.W.; Han, S.Y.; Nguyen, M.C.; Nguyen, A.H.T.; Kim, J.Y.; Choi, S.; Cheon, J.; Ji, H.; Choi, R. Low-Temperature Solution-Based In_2O_3 Channel Formation for Thin-Film Transistors Using a Visible Laser-Assisted Combustion Process. *IEEE Electron. Device Lett.* **2017**, *38*, 1259–1262. [CrossRef]
67. Noh, M.; Seo, I.; Park, J.; Chung, J.S.; Lee, Y.S.; Kim, H.J.; Chang, Y.J.; Park, J.H.; Kang, M.G.; Kang, C.Y. Spectroscopic ellipsometry investigation on the excimer laser annealed indium thin oxide sol-gel films. *Curr. Appl. Phys.* **2016**, *16*, 145–149. [CrossRef]
68. Uraoka, Y.; Corsino, D.; Bermundo, J.P.; Fujii, M.N.; Uenuma, M. Highly Reliable Metal Oxide Thin Film Transistors for Flexible Devices. *ECS Trans.* **2020**, *98*, 29–37. [CrossRef]
69. Dellis, S.; Isakov, I.; Kalfagiannis, N.; Tetzner, K.; Anthopoulos, T.D.; Koutsogeorgis, D.C. Rapid laser-induced photochemical conversion of sol-gel precursors to In_2O_3 layers and their application in thin-film transistors. *J. Mater. Chem. C* **2017**, *5*, 3673–3677. [CrossRef]
70. Tsuchiya, T.; Yamaguchi, F.; Morimoto, I.; Nakajima, T.; Kumagai, T. Microstructure control of low-resistivity tin-doped indium oxide films grown by photoreaction of nanoparticles using a KrF excimer laser at room temperature. *Appl. Phys. A-Mater. Sci. Process.* **2010**, *99*, 745–749. [CrossRef]
71. Nian, Q.; Callahan, M.; Saei, M.; Look, D.; Efstathiadis, H.; Bailey, J.; Cheng, G.J. Large Scale Laser Crystallization of Solution-based Alumina-doped Zinc Oxide (AZO) Nanoinks for Highly Transparent Conductive Electrode. *Sci Rep.* **2015**, *5*, 12. [CrossRef]
72. Pan, H.; Misra, N.; Ko, S.H.; Grigoropoulos, C.P.; Miller, N.; Haller, E.E.; Dubon, O. Melt-mediated coalescence of solution-deposited ZnO nanoparticles by excimer laser annealing for thin-film transistor fabrication. *Appl. Phys. A-Mater. Sci. Process.* **2009**, *94*, 111–115. [CrossRef]
73. Lee, D.; Pan, H.; Ko, S.H.; Park, H.K.; Kim, E.; Grigoropoulos, C.P. Non-vacuum, single-step conductive transparent ZnO patterning by ultra-short pulsed laser annealing of solution-deposited nanoparticles. *Appl. Phys. A-Mater. Sci. Process.* **2012**, *107*, 161–171. [CrossRef]
74. Yang, Y.H.; Yang, S.S.; Chou, K.S. Characteristic Enhancement of Solution-Processed In-Ga-Zn Oxide Thin-Film Transistors by Laser Annealing. *IEEE Electron. Device Lett.* **2010**, *31*, 969–971. [CrossRef]
75. Chen, C.H.; Yang, H.H.; Yang, Q.; Chen, G.X.; Chen, H.P.; Guoi, T.L. Low-temperature solution-processed flexible metal oxide thin-film transistors via laser annealing. *J. Phys. D-Appl. Phys.* **2019**, *52*, 11. [CrossRef]

76. Huang, H.; Hu, H.L.; Zhu, J.G.; Guo, T.L. Inkjet-Printed In-Ga-Zn Oxide Thin-Film Transistors with Laser Spike Annealing. *J. Electron. Mater.* **2017**, *46*, 4497–4502. [CrossRef]
77. Ning, H.-L.; Deng, Y.-X.; Huang, J.-L.; Luo, Z.-L.; Hu, R.-D.; Liu, X.-Z.; Wang, Y.-P.; Qiu, T.; Yao, R.-H.; Peng, J.-B. Laser annealing of metal oxide thin film transistor. *Chin. J. Liq. Cryst. Disp.* **2020**, *35*, 1211–1221. [CrossRef]
78. Bradford, J.N.; Williams, R.T.; Faust, W.L. Study of F-center formation in KCl on a picosecond time scale. *Phys. Rev. Lett.* **1975**, *35*, 300–303. [CrossRef]
79. Young, R.T.; Narayan, J.; Christie, W.H.; van der Leeden, G.A.; Levatter, J.I.; Cheng, L.J. Semiconductor processing with excimer lasers. *Solid State Technol.* **1983**, *26*, 183–189.
80. Sandu, C.S.; Teodorescu, V.S.; Ghica, C.; Hoffmann, P.; Bret, T.; Brioude, A.; Blanchin, M.G.; Roger, J.A.; Canut, B.; Croitoru, M. Excimer laser crystallization of SnO_2:Sb sol-gel films. *J. Sol.-Gel Sci. Technol.* **2003**, *28*, 227–234. [CrossRef]
81. Sandu, C.S.; Teodorescu, V.S.; Ghica, C.; Canut, B.; Blanchin, M.G.; Roger, J.A.; Brioude, A.; Bret, T.; Hoffmann, P.; Garapon, C. Densification and crystallization of SnO_2:Sb sol-gel films using excimer laser annealing. *Appl. Surf. Sci.* **2003**, *208*, 382–387. [CrossRef]
82. Ding, L.; Nicolay, S.; Steinhauser, J.; Kroll, U.; Ballif, C. Relaxing the Conductivity/Transparency Trade-Off in MOCVD ZnO Thin Films by Hydrogen Plasma. *Adv. Funct. Mater.* **2013**, *23*, 5177–5182. [CrossRef]
83. Ellmer, K.; Klein, A.; Rech, B. *Transparent Conductive Zinc Oxide*; Springer: Berlin, Germany, 2008.
84. Antonio, L.; Steven, H. *Handbook of Photovoltaic Science and Engineering*; John Wiley & Sons, Ltd.: Hoboken, NJ, USA, 2010.
85. Minami, T. New n-type transparent conducting oxides. *MRS Bull.* **2000**, *25*, 38–44. [CrossRef]
86. Nagase, T.; Ooie, T.; Sakakibara, J. A novel approach to prepare zinc oxide films: Excimer laser irradiation of sol-gel derived precursor films. *Thin Solid Films* **1999**, *357*, 151–158. [CrossRef]
87. Voutsas, A.T.; Marmorstein, A.M.; Solanki, R. The impact of annealing ambient on the performance of excimer-laser-annealed polysilicon thin-film transistors. *J. Electrochem. Soc.* **1999**, *146*, 3500–3505. [CrossRef]
88. Ozerov, I.; Nelson, D.; Bulgakov, A.V.; Marine, W.; Sentis, M. Synthesis and laser processing of ZnO nanocrystalline thin films. *Appl. Surf. Sci.* **2003**, *212*, 349–352. [CrossRef]
89. Weidenkaff, A.; Steinfeld, A.; Wokaun, A.; Auer, P.O.; Eichler, B.; Reller, A. Direct solar thermal dissociation of zinc oxide: Condensation and crystallisation of zinc in the presence of oxygen. *Sol. Energy* **1999**, *65*, 59–69. [CrossRef]
90. Janotti, A.; Van de Walle, C.G. Oxygen vacancies in ZnO. *Appl. Phys. Lett.* **2005**, *87*, 3. [CrossRef]
91. Corsino, D.C.; Bermundo, J.P.S.; Kulchaisit, C.; Fujii, M.N.; Ishikawa, Y.; Ikenoue, H.; Uraoka, Y. High-Performance Fully Solution-Processed Oxide Thin-Film Transistors via Photo-Assisted Role Tuning of InZnO. *ACS Appl. Electron. Mater.* **2020**, *2*, 2398–2407. [CrossRef]
92. Huang, H.Y.; Wang, S.J.; Wu, C.H.; Lu, C.Y. Improvement of Electrical Performance of InGaZnO/HfSiO TFTs with 248-nm Excimer Laser Annealing. *Electron. Mater. Lett.* **2014**, *10*, 899–902. [CrossRef]
93. Socratous, J.; Banger, K.K.; Vaynzof, Y.; Sadhanala, A.; Brown, A.D.; Sepe, A.; Steiner, U.; Sirringhaus, H. Electronic Structure of Low-Temperature Solution-Processed Amorphous Metal Oxide Semiconductors for Thin-Film Transistor Applications. *Adv. Funct. Mater.* **2015**, *25*, 1873–1885. [CrossRef] [PubMed]
94. Hwang, S.; Lee, J.H.; Woo, C.H.; Lee, J.Y.; Cho, H.K. Effect of annealing temperature on the electrical performances of solution-processed InGaZnO thin film transistors. *Thin Solid Films* **2011**, *519*, 5146–5149. [CrossRef]
95. Huang, X.M.; Wu, C.F.; Lu, H.; Ren, F.F.; Chen, D.J.; Liu, Y.L.; Yu, G.; Zhang, R.; Zheng, Y.D.; Wang, Y.J. Large-Swing a-IGZO Inverter with a Depletion Load Induced by Laser Annealing. *IEEE Electron. Device Lett.* **2014**, *35*, 1034–1036. [CrossRef]
96. Bäuerle, D. *Laser Processing and Chemistry*; Springer: Berlin, Germany, 2011.
97. Narendra, B.D.; Sandip, P.H. *Laser Fabrication and Machining of Materials*; Springer: Berlin, Germany, 2008.
98. Palneedi, H.; Maurya, D.; Kim, G.Y.; Annapureddy, V.; Noh, M.S.; Kang, C.Y.; Kim, J.W.; Choi, J.J.; Choi, S.Y.; Chung, S.Y.; et al. Unleashing the Full Potential of Magnetoelectric Coupling in Film Heterostructures. *Adv. Mater.* **2017**, *29*, 9. [CrossRef] [PubMed]
99. Bogdan Allemann, I.; Kaufman, J. Laser principles. *Curr. Probl. Dermatol.* **2011**, *42*, 7–23. [CrossRef] [PubMed]
100. Gamaly, E.G.; Rode, A.V. Physics of ultra-short laser interaction with matter: From phonon excitation to ultimate transformations. *Prog. Quantum Electron.* **2013**, *37*, 215–323. [CrossRef]

Article

Atmosphere Effect in Post-Annealing Treatments for Amorphous InGaZnO Thin-Film Transistors with SiO_x Passivation Layers

Wen Zhang, Zenghui Fan, Ao Shen and Chengyuan Dong *

Department of Electronic Engineering, Shanghai Jiao Tong University, Shanghai 200240, China; wen_zhang0032@sjtu.edu.cn (W.Z.); zenghui_f@sjtu.edu.cn (Z.F.); shenao1997@sjtu.edu.cn (A.S.)
* Correspondence: cydong@sjtu.edu.cn; Tel.: +86-21-3420-8271

Abstract: We investigated the electrical performance and positive bias stress (PBS) stability of the amorphous InGaZnO thin-film transistors (a-IGZO TFTs) with SiO_x passivation layers after the post-annealing treatments in different atmospheres (air, N_2, O_2 and vacuum). Both the chamber atmospheres and the device passivation layers proved important for the post-annealing effects on a-IGZO TFTs. For the heat treatments in O_2 or air, the larger threshold voltage (V_{TH}) and off current (I_{OFF}), smaller field-effect mobility (μ_{FE}), and slightly better PBS stability of a-IGZO TFTs were obtained. The X-ray photoemission spectroscopy (XPS) and secondary ion mass spectroscopy (SIMS) measurement results indicated that the oxygen atoms from the ambience led to less oxygen vacancies (V_O) and more oxygen-related defects in a-IGZO after the heat treatments in O_2 or air. For the annealing processes in vacuum or N_2, the electrical performance of the a-IGZO TFTs showed nearly no change, but their PBS stability evidently improved. After 4500 seconds' stressing at 40 V, the V_{TH} shift decreased to nearly 1 V. In this situation, the SiO_x passivation layers were assumed to effectively prevent the oxygen diffusion, keep the V_O concentration unchanged and refuse the oxygen-related defects into the a-IGZO films.

Keywords: amorphous InGaZnO (a-IGZO); thin-film transistor (TFT); positive bias stress (PBS); annealing atmosphere; oxygen vacancy

Citation: Zhang, W.; Fan, Z.; Shen, A.; Dong, C. Atmosphere Effect in Post-Annealing Treatments for Amorphous InGaZnO Thin-Film Transistors with SiO_x Passivation Layers. *Micromachines* **2021**, *12*, 1551. https://doi.org/10.3390/mi12121551

Academic Editor: Nam-Trung Nguyen

Received: 1 December 2021
Accepted: 9 December 2021
Published: 12 December 2021

1. Introduction

Amorphous InGaZnO thin-film transistors (a-IGZO TFTs) have been regarded as among the most promising active-matrix devices for next-generation flat panel displays (FPDs) due to their high mobility, good uniformity and low fabrication temperature [1–5]. For mass productions of a-IGZO TFTs, either etching-stopper (ES) or back-channel-etching (BCE) structures are used. Although the better electrical performance and stability are obtained for ES-structured devices, BCE is still preferred for its simpler processing and compatibility with amorphous silicon (a-Si) TFT productions [6–9]. Hence, researchers have been taking many measures to improve the electrical performance and stability of BCE-structured a-IGZO TFTs, among which post-annealing treatments seem extremely useful [10,11]. Some studies pointed out that the annealing effect was closely associated with the oxygen vacancy (V_O), one of the most important defects in a-IGZO films [12,13]. Assuming V_O was sensitive to annealing atmospheres, as some researchers obtained higher performances and stabler properties of a-IGZO TFTs by choosing appropriate annealing atmospheres (e.g., oxygen, nitrogen, vacuum, and so on) [14–17]. However, all of these studies were limited to the unpassivated devices whereas the TFT devices are always passivated in mass production. Since passivation layers might exhibit different effects under different annealing atmospheres, it might be meaningful to investigate the atmosphere effect in post-annealing treatments for the a-IGZO TFTs with passivation ayers.

In this study, the electrical performance and positive bias stress (PBS) stability of the a-IGZO TFTs with SiO_x passivation layers under various post-annealing atmospheres (air,

O_2, N_2 and vacuum) were comparatively investigated, whose physical essence was deeply analyzed with the help of X-ray photoemission spectroscopy (XPS) and secondary ion mass spectroscopy (SIMS). We found that both the chamber atmospheres and the device passivation layers played important roles in the post-annealing treatments for a-IGZO TFTs. A qualitative model was also built to explain the experimental results.

2. Materials and Methods

Figure 1 shows the schematic cross-section of the BCE-structured a-IGZO TFTs used in this study. Firstly, the gate electrodes (150 nm-thick aluminum and 23 nm-thick molybdenum films) were deposited on the glass substrates by sputtering. Then, 350 nm-thick SiN_x and 50 nm-thick SiO_x films were prepared as gate insulators (GIs) by plasma-enhanced chemical vapor deposition (PECVD). The 60 nm-thick channel layers were deposited by RF sputtering at room temperature (RT) using an IGZO target, where the atoms ratio of In:Ga:Zn was 1:1:1. After patterning the channel layers, a pre-annealing treatment (300 °C/1 h/air) was employed in a furnace, which could increase the film density/uniformity and reduce the defects in the a-IGZO films. The 30 nm-thick molybdenum and 150 nm-thick aluminum films were then prepared and patterned as source/drain (S/D) electrodes before the 200 nm-thick SiO_x films were deposited as passivation layers by PECVD. Finally, the samples were annealed at 300 °C for 1 h in different atmospheres (air, O_2, N_2 and vacuum) using a rapid thermal processing (RTP) oven, where the gas flows of O_2 (or N_2) were set to 1 L/min and the vacuum pressure was kept at 6.5×10^{-3} Pa. It is worth noting that all the films were patterned by standard lithography and etching processes. The typical channel width/length of the a-IGZO TFTs was 12/6 μm.

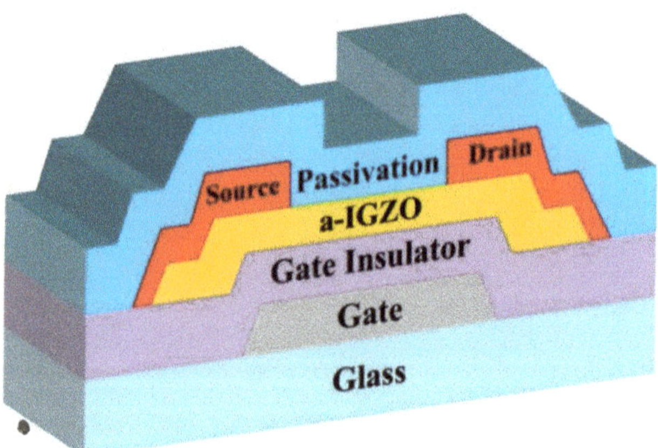

Figure 1. Schematic cross-section of the back-channel-etching (BCE) -structured amorphous InGaZnO thin-film transistors (a-IGZO TFTs) in this study.

The electrical performance and PBS stability of the a-IGZO TFTs were measured by a Keithley 2636 analyzer. The transfer curves were obtained when V_{GS} was scanned from −20 V to 40 V at a step of 0.5 V and V_{DS} was set to 10 V. During the PBS tests, the transfer curves were instantly measured following each 1500 s' stressing (V_{GS} = +40 V). All the above tests were performed in a dark chamber at RT. An XPS analyzer (AXIS Ultra DLD) was used to characterize the chemical bonding states of the a-IGZO films (without or with passivation layers) annealed under different atmospheres. Here, an X-ray source with the aluminum anode was used to bombard the film surface and obtain the energy spectrum of different elements. The binding energy of all elements were calibrated using the C1s peak at 284.8 eV. A SIMS measurement system (TOF SIMS 5 produced by ION-TOF GmbH) with a Cs^+ primary ion source was applied to measure the depth profiles of different elements.

3. Results and Discussion

Figure 2a shows the transfer curves of the a-IGZO TFTs annealed under different atmospheres, where the unannealed device was considered the reference sample. The corresponding performance parameters including threshold voltage (V_{TH}), field-effect mobility (μ_{FE}), sub-threshold swing (SS) and on–off current ratio (I_{ON}/I_{OFF}) were extracted from the transfer curves according to Ref. [18]. Here, 6–12 devices were measured and statistically analyzed. As shown in Figure 2b, the average mobilities of the devices annealed under different atmospheres (unannealed, air, O_2, N_2 and vacuum) were 3.07, 2.08, 1.37, 3.78 and 4.03 $cm^2/(V{\cdot}s)$, respectively. One may notice that the annealing in O_2 or air obviously degraded the device mobility, whereas the annealing in N_2 or vacuum slightly improved μ_{FE}. In addition, the corresponding V_{TH} values were 3.6, 5.1, 8.0, 4.7 and 5.4 V, the SS values were 2.8, 2.4, 2.7, 2.6 and 2.4 V/dec, and the on–off current ratios were 3.8×10^6, 2.4×10^6, 3.9×10^5, 8.2×10^6 and 1.3×10^7, respectively (see Figure 2b,c). Compared with the unannealed sample, one may observe from Figure 2b that the post-annealing in O_2 evidently increased the threshold voltages of a-IGZO TFTs, while the annealing treatments in N_2 or vacuum showed nearly no influence on the V_{TH} values. As shown in Figure 2c, the SS values of a-IGZO TFTs remained almost the same under various post-annealing atmospheres. The post-annealing treatments resulted in sufficiently large on/off current ratios (>10^6) except that in O_2, as shown in Figure 2c.

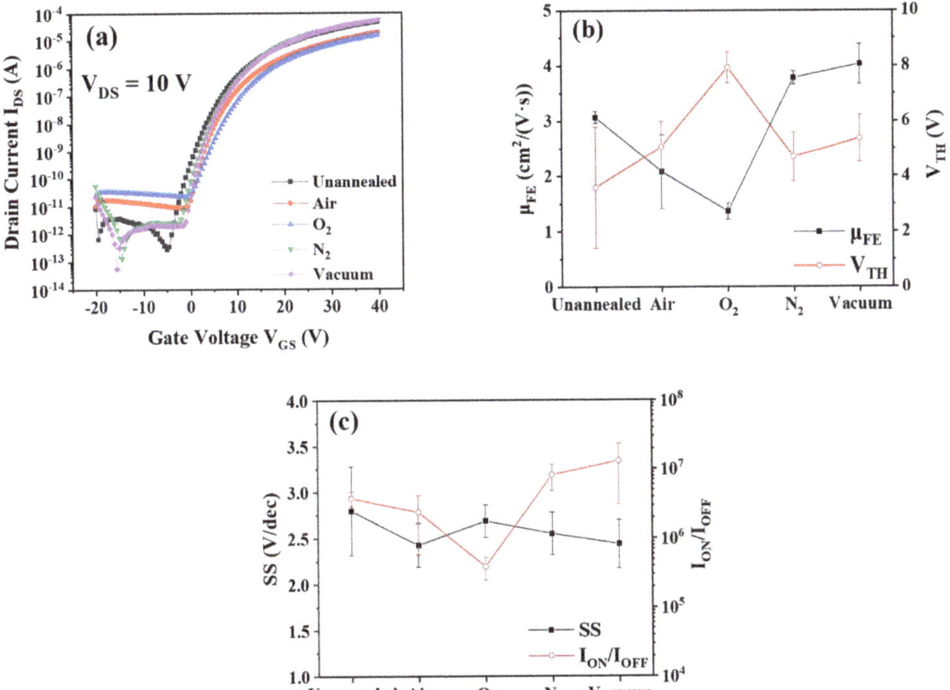

Figure 2. (**a**) Transfer curves; (**b**) mobility/threshold voltage; and (**c**) sub-threshold slope/on–off current ratio of the a-IGZO TFTs with SiO_x passivation layers annealed in different atmospheres.

The bias stress effect plays an important role in the actual applications of TFT devices [2,19]. Accordingly, we measured the PBS stability of the a-IGZO TFTs with SiO_x passivation under various annealing atmospheres (unannealed, air, O_2, N_2 and vacuum). Figure 3a shows the transfer curve evolution of the unannealed a-IGZO TFTs during PBS tests. With the stressing time elapsed, the transfer curves of the devices shifted positively.

In order to quantitatively describe this unstable property, we defined a useful term ΔV_{TH} (the V_{TH} difference between the after-stressing state and the initial state) in this study. It is worth noting that each ΔV_{TH} value was averaged over 3–6 devices. After 4500 s' stressing, ΔV_{TH} of the unannealed device became 3.67 V. All the post-annealed devices exhibited better bias stress stability than the unannealed sample. It is interesting that the ΔV_{TH} value distinctively depended on the post-annealing atmosphere. As shown in Figure 3b, ΔV_{TH} decreased to 1.88 V, 2.50 V, 1.13 V and 1.38 V for the devices annealed in air, O_2, N_2 and vacuum, respectively. It is worth noting that the post-annealing treatments in N_2 and vacuum showed much larger improvements than those in O_2 and air.

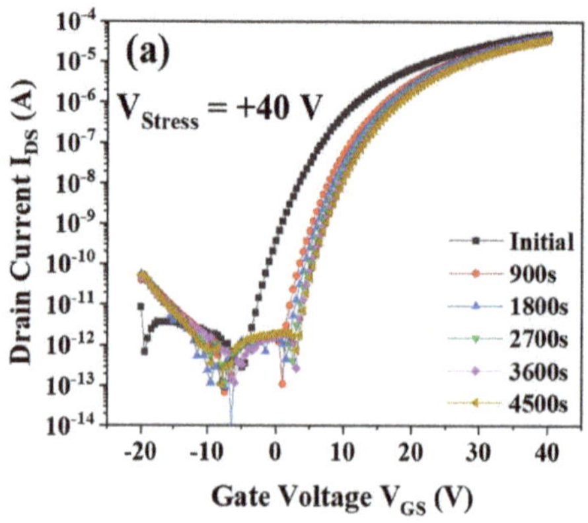

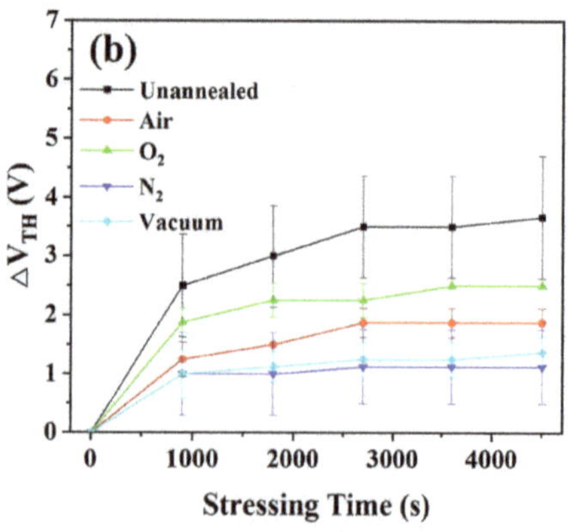

Figure 3. (**a**) Stressing time dependence of the transfer curves of the unannealed a-IGZO TFTs with SiO_x passivation layers; (**b**) threshold voltage shifts of the a-IGZO TFTs with SiO_x passivation layers annealed under different atmospheres during PBS tests.

In order to determine the related physical mechanisms, we used the XPS technique to analyze the a-IGZO films (without or with passivation layers) annealed under various atmospheres. Firstly, we deposited 60 nm-thick a-IGZO films on glass substrates, annealed them under different atmospheres, and applied them to XPS measurements. Although these unpassivated films were different from the real situation in the TFT devices (see Figure 1), they could more evidently exhibit the influences of the annealing atmospheres on the back surfaces of the a-IGZO films. Figure 4a shows the deconvolution diagram of the O1s spectrum of the unannealed sample. The O1s peak can be deconvoluted into three sub-peaks using Gaussian fitting method which are approximately centered at 529.5 eV (O_I), 530.9 eV (O_{II}) and 532.0 eV (O_{III}), respectively. The O_I peak represents the oxygen bonds with metal, the O_{II} peak is bound up with V_O, and the O_{III} peak is related to the hydrated oxides defects [20,21]. Figure 4b,c show the peak area ratios of the O_{II} and O_{III} over the total area of O1s peak ($O_{Total} = O_I + O_{II} + O_{III}$). The area ratio O_{II}/O_{Total} of the a-IGZO films annealed under various atmospheres (unannealed, air, O_2, N_2 and vacuum) were 28.56%, 28.19%, 21.78%, 30.20% and 30.02%, respectively. One may observe that the post-annealing in O_2 largely decreased the area ratio O_{II}/O_{Total}, whereas the heat treatments in air/N_2/vacuum showed little effect in this regard. On the other side, the area ratio O_{III}/O_{Total} of the a-IGZO films under different atmospheres (unannealed, air, O_2, N_2 and vacuum) were 13.08%, 16.91%, 17.19%, 17.53% and 17.37%, respectively. It is obvious that all the post-annealing treatments increased the area ratio O_{III}/O_{Total}.

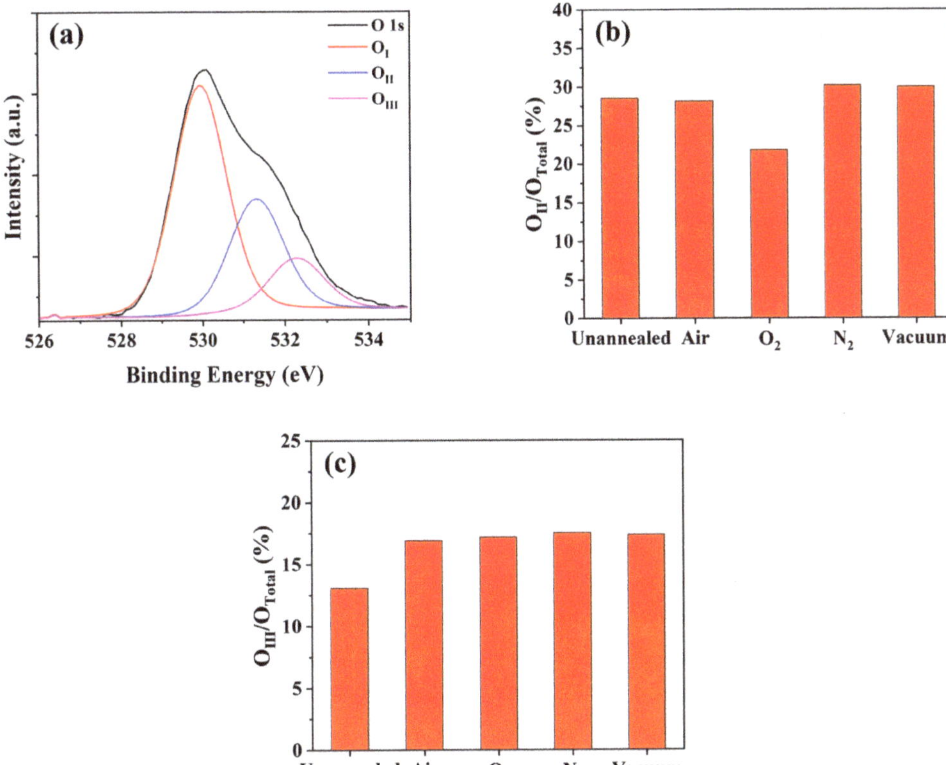

Figure 4. (**a**) Deconvolution diagram of the O1s peak of the unannealed a-IGZO films; the area ratios of (**b**) O_{II} and (**c**) O_{III} in the O1s peak of the a-IGZO films annealed under different atmospheres.

As shown in Figure 4b, only O_2_annealing among five treatments evidently changed V_O in a-IGZO films, which are in need of much further study. Most importantly, could passivation layers prevent this effect? Attempting to address this question, we used XPS depth profiling technique to analyze the chemical bonding states in the a-IGZO films passivated by 50 nm-thick SiO_x [22]. As shown in Figure 5a, point A was exactly under the interface between the a-IGZO and SiO_x, whereas point B was in the middle of a-IGZO films. The characterization results of the unannealed samples and the O_2_annealed samples are shown in Figure 5b. At point A, the area ratio O_{II}/O_{Total} decreased from 31.0% to 25.8% by the O_2_annealing treatment, which was quite similar to the case of the unpassivated a-IGZO films (see Figure 4b). At point B, however, there was nearly no difference in the area ratio O_{II}/O_{Total} between the unannealed (34.5%) and O_2_annealed (34.8%) samples. These results indicated that the oxygen atoms might also diffuse into the a-IGZO back surfaces and hence combine with V_O even if the SiO_x passivation layer was applied during the O_2_annealing treatments. Although the thickness of SiO_x used here (50 nm) was thinner than that of the passivation layer (200 nm) in the a-IGZO TFTs, we assume that this conclusion was also applied to the real devices.

In order to further ascertain the role of the ambient oxygen atoms during the thermal annealing treatments, we comparatively measured the SIMS profiles of the unannealed sample and O_2_annealed sample, both of which consisted of 60 nm-thick a-IGZO and 160 nm-thick SiO_x films. Here, the SIMS depth profiling technique was used to analyze the distribution of different elements in the unannealed and O_2-annealed samples [23]. Figure 6a shows the depth dependence of the second ion intensity for the unannealed sample. One can observe that there were abrupt changes in Si, InO, GaO and ZnO at the depth of approximately 160 nm, which was close to the interfacial surface between SiO_x and a-IGZO. The O_2_annealing brought nearly the same SIMS picture as that of the unannealed sample except the data about oxygen ions. As shown in Figure 6b, the secondary ion intensity of oxygen remained almost unchanged within the SiO_x films by the O_2_annealing, implying that the entering oxygen atoms from the ambience during the annealing treatment in O_2 were too few to be detected by SIMS. Therefore, we reasonably assume that the oxygen atoms entering the a-IGZO also could not be detected by SIMS. However, as shown in Figure 6b, the secondary ion intensity of the oxygen apparently increased for the O_2_annealed sample. We assume that the microstructure of the a-IGZO evidently changed during the annealing treatment in O_2, which might be due to the entering oxygen atoms from the ambience although their concentration was quite small. This change also implied that the O_2_annealing might bring some oxygen-related-defects such as interstitial oxygen in the a-IGZO films [24,25].

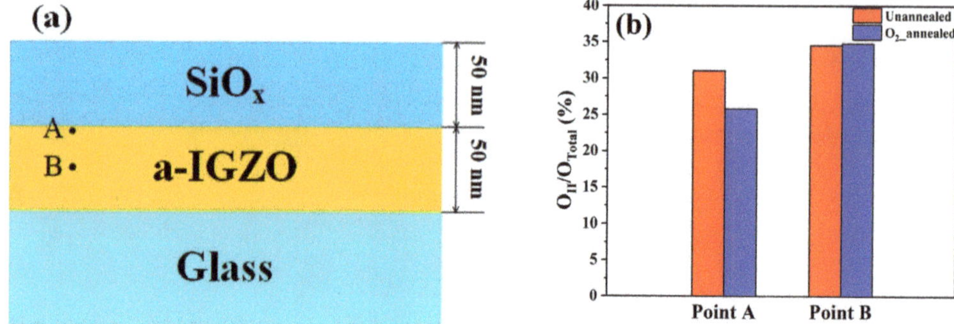

Figure 5. (**a**) The sample structure for the XPS depth profiling tests; (**b**) the area ratios of O_{II} in the O1s peak of the unannealed and O_2_annealed samples.

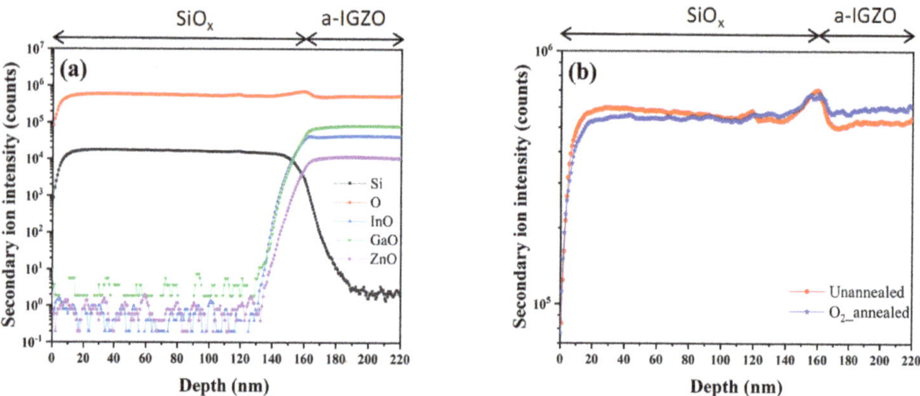

Figure 6. (**a**) SIMS depth profiles for the unannealed sample; (**b**) the oxygen depth distributions for the unannealed sample and the O_2_annealed sample.

It is well known that a-IGZO films are very sensitive to ambient atmospheres, especially oxygen and water [26,27]. For mass production, passivation layers are always used to isolate a-IGZO TFTs from the ambience. However, post-annealing treatments at high temperatures accelerate the interaction between the a-IGZO and outside world. This interaction is a dynamic process during annealing treatments. Here, we limit our discussion to the interactions between oxygen atoms because they are dominant in the electrical performance and stable properties of a-IGZO TFTs. Figure 7a shows the schematic diagram of the dynamic process during the annealing treatments at 300 °C in O_2 (or air) atmospheres. In this situation, the oxygen atoms were full of the RTP chamber, so the oxygen diffusion from the chamber into the a-IGZO films should prevail that from the device channel layers into the ambience. It was reported that oxygen atoms could pass through dielectric films due to thermal diffusion at high temperatures [28,29]. Importantly, these newcomers (oxygen atoms) combined with oxygen vacancies and effectively decreased the V_O concentration in a-IGZO films, which was confirmed by the XPS measurement results (see Figures 4b and 5b). This drop in the V_O concentration accordingly resulted in the larger V_{TH} values for the corresponding TFT devices, as shown in Figure 2b. It is worth noting that this effect was evidently weakened for the annealing treatments in air due to the existence of much more nitrogen atoms in the RTP chamber. Furthermore, the entering oxygen atoms also brought some oxygen-related-defects (see the discussion about the SIMS measurement results), and led to smaller μ_{FE} and larger I_{OFF} (as shown in Figure 2). Generally, heat treatments could improve the bias stress stability of a-IGZO TFTs due to the drops in the bulk defects and interface traps [30–32]; however, these oxygen-related-defects degraded this improvement effect. Specifically, the oxygen-related defects might make the V_O formation easier during positive bias stressing [33], which resulted in more unstable properties of a-IGZO TFTs. Thus, the PBS stability improvements for the a-IGZO TFTs annealed in O_2 and air were not so evident, as shown in Figure 3b.

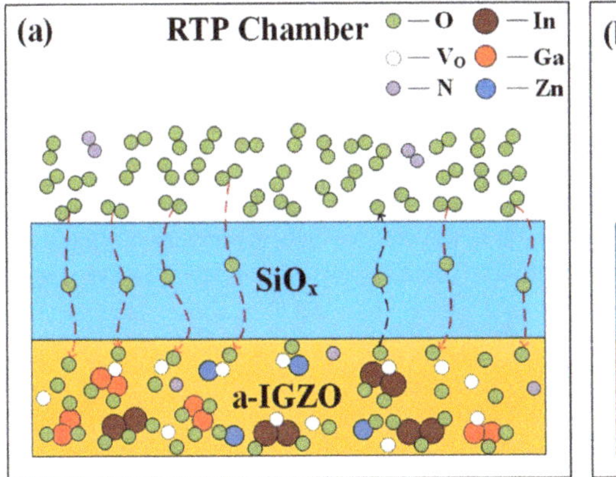

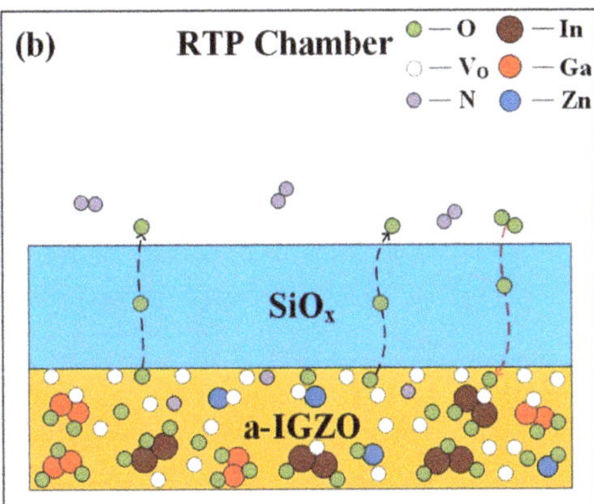

Figure 7. Schematic diagrams of the dynamic processes during the post-annealing treatments at 300 °C in (**a**) O_2 (or air) and (**b**) vacuum (or N_2).

Figure 7b shows the dynamic process during the post-annealing at 300 °C in vacuum (or N_2). In this situation, there were few oxygen atoms in the RTP chamber, so the oxygen diffusion from the a-IGZO film to the ambience was assumed to be dominant. If so, the V_O concentration would have become larger. However, this rise in the V_O concentration was not apparent due to the XPS measurement data. As shown in Figure 4b, the annealing in N_2 (or vacuum) hardly changed the area ratio O_{II}/O_{Total}. We ascribe this phenomenon to the barrier effect of the SiO_x passivation layers. In contrast to Figure 7a, the SiO_x passivation layers effectively prevented the oxygen diffusion from the a-IGZO into the ambience during heat treatments in vacuum (or N_2). Since the V_O concentration changed little in this case, the electron concentration in the channel layers of the a-IGZO TFTs remained nearly unchanged, leading to V_{TH} and μ_{FE} values similar to those of the unannealed device (see Figure 2b). As for the bias stress stability of the a-IGZO TFTs annealed in vacuum (or N_2), we reasonably assume that there were no oxygen-related-defects included here, so the improvement effect over the stable properties of a-IGZO TFTs by heat treatments could be well kept [34]. As shown in Figure 3b, the annealing in vacuum (or N_2) effectively improved the PBS stability of the a-IGZO TFTs with SiO_x passivation layers, which should be preferred in mass productions.

Finally, we briefly discussed two interesting questions relating this study. The first one is about the function of V_O and the other one is relating the choice of passivation materials. Some researchers reported the V_O drop led to better device performance [35,36], which is different from the results here. We ascribe this difference to the special effects of O_2_annealing treatments, i.e., they not only decreased the V_O concentration but also brought some oxygen-related-defects [24,25]. In this study, the SiO_x was selected as passivation layers, which was found to play important roles in the post-annealing treatments under different atmospheres. We assumed that other passivation materials should bring similar tendencies but to different extents. This assumption might be supported by the other research results from our group, which indicated that the a-IGZO TFTs with SiO_x and AlO_x passivation layers exhibited almost the same PBS stability if the different passivation layer thicknesses were used [33]. In fact, not only the passivation material but also some treatments on passivation layers might also influence their performance during heat treatments. It is reported that the thermal stability of TFTs can be improved by using fluorinated organic passivation, where the diffusion of F during the annealing process reduced the V_O, leading to better thermal stability [35]. Although the different passivation layers showed somewhat different performances, all the above reports suggested that passivation layers have similar barrier effects during post-annealing treatments.

4. Conclusions

The post-annealing atmosphere effectively influenced the electrical performance and bias stress stability of the a-IGZO TFTs with SiO_x passivation layers. For the heat treatments in O_2 (or air), the oxygen atoms in the ambience diffused into the a-IGZO and combined with oxygen vacancies, leading to less V_O and more oxygen-related defects; this resulted in larger V_{TH}/I_{OFF}, smaller μ_{FE} and the slightly better PBS stability of the a-IGZO TFTs. For the post-annealing process in vacuum (or N_2), the oxygen diffusion was effectively prevented by the SiO_x passivation layers, and hence the V_O concentration hardly changed; the electrical performance of the a-IGZO TFTs showed nearly no change, but very importantly, their PBS stability evidently improved. This qualitative model was confirmed by the XPS and SIMS measurement results.

Author Contributions: Conceptualization, W.Z., Z.F., A.S. and C.D.; investigation, W.Z.; writing, W.Z., Z.F., A.S. and C.D. All authors have read and agreed to the published version of the manuscript.

Funding: This work was supported by the Key Research Project of Jiangxi Province (Grant No. 20194ABC28005). The authors thank Chen Wang, Cong Peng and Xi Feng Li (Shanghai University) for their support in the device fabrications.

Data Availability Statement: The data presented in this study are available on request from the corresponding author.

Conflicts of Interest: The authors declare no conflict of interest.

References

1. Nomura, K.; Ohta, H.; Takagi, A.; Kamiya, T.; Hirano, M.; Hosono, H. Room-temperature fabrication of transparent flexible thin-film transistors using amorphous oxide semiconductors. *Nature* **2004**, *432*, 488–492. [CrossRef]
2. Kamiya, T.; Nomura, K.; Hosono, H. Present status of amorphous In–Ga–Zn–O thin-film transistors. *Sci. Technol. Adv. Mater.* **2010**, *11*, 044305. [CrossRef]
3. Yabuta, H.; Sano, M.; Abe, K.; Aiba, T.; Den, T.; Kumomi, H.; Nomura, K.; Kamiya, T.; Hosono, H. High-mobility thin-film transistor with amorphous InGaZnO4 channel fabricated by room temperature rf-magnetron sputtering. *Appl. Phys. Lett.* **2006**, *89*, 112123. [CrossRef]
4. Lee, J.; Kim, D.; Yang, D.; Hong, S.; Yoon, K.; Hong, P.; Jeong, C.; Park, H.; Kim, S.; Lim, S.; et al. 42.2: World's largest (15-inch) XGA AMLCD panel using IGZO oxide TFT. *SID Symp. Dig. Tech. Pap.* **2008**, *39*, 625–628. [CrossRef]
5. Jeong, J.; Jeong, J.; Choi, J.; Im, J.; Kim, S.; Yang, H.; Kang, K.; Kim, K.; Ahn, T.; Chung, H.; et al. 3.1: Distinguished Paper: 12.1-Inch WXGA AMOLED display driven by indium-gallium-zinc oxide TFTs array. *SID Symp. Dig. Tech. Pap.* **2008**, *39*, 1–4. [CrossRef]

6. Kim, M.K.; Jeong, J.H.; Lee, H.J.; Ahn, T.K.; Shin, H.S.; Park, J.S.; Jeong, J.K.; Mo, Y.G.; Kim, H.D. High mobility bottom gate InGaZnO thin film transistors with SiOx etch stopper. *Appl. Phys. Lett.* **2007**, *90*, 212114. [CrossRef]
7. Peng, C.; Yang, S.; Pan, C.; Li, X.; Zhang, J. Effect of two-step annealing on high stability of a-IGZO thin-film transistor. *IEEE Trans. Electron Devices* **2020**, *67*, 4262–4268. [CrossRef]
8. Nag, M.; Bhoolokam, A.; Steudel, S.; Chasin, A.; Myny, K.; Maas, J.; Groeseneken, G.; Heremans, P. Back-channel-etch amorphous indium–gallium–zinc oxide thin-film transistors: The impact of source/drain metal etch and final passivation. *Jpn. J. Appl. Phys.* **2014**, *53*, 111401. [CrossRef]
9. Park, Y.; Um, J.; Mativenga, M.; Jang, J. Enhanced operation of back-channel-etched a-IGZO TFTs by fluorine treatment during source/drain wet-etching. *ECS J. Solid State Sci. Technol.* **2017**, *6*, 300–303. [CrossRef]
10. Ochi, M.; Hino, A.; Goto, H.; Hayashi, K.; Kugimiya, T. Improvement of stress stability in back channel etch-type amorphous In–Ga–Zn–O thin film transistors with post process annealing. *ECS J. Solid State Sci. Technol.* **2017**, *6*, 247–252. [CrossRef]
11. Yu, Y.; Lv, N.; Zhang, D.; Wei, Y.; Wang, M. High-mobility amorphous InGaZnO thin-film transistors with nitrogen introduced via low-temperature annealing. *IEEE Electron Device Lett.* **2021**, *42*, 1480–1483. [CrossRef]
12. Trinh, T.; Nguyen, V.; Ryu, K.; Jang, K.; Lee, W.; Baek, S.; Raja, J.; Yi, J. Improvement in the performance of an InGaZnO thin-film transistor by controlling interface trap densities between the insulator and active layer. *Semicond. Sci. Technol.* **2011**, *26*, 085012. [CrossRef]
13. Jung, C.; Kim, D.; Kang, Y.; Yoon, D. Effect of heat treatment on electrical properties of amorphous oxide semiconductor In–Ga–Zn–O film as a function of oxygen flow rate. *Jpn. J. Appl. Phys.* **2009**, *48*, 08HK02. [CrossRef]
14. Lin, C.; Yen, T.; Lin, H.; Huang, T.; Lee, Y. Effect of annealing ambient on the characteristics of a-IGZO thin film transistors. In Proceedings of the 4th IEEE International NanoElectronics Conference, Taiwan, China, 21–24 June 2011; pp. 1–2.
15. Fuh, C.; Sze, S.; Liu, P.; Teng, L.; Chou, Y. Role of environmental and annealing conditions on the passivation-free In–Ga–Zn–O TFT. *Thin Solid Films* **2011**, *520*, 1489–1494. [CrossRef]
16. Gutierrez-Heredia, G.; Maeng, J.; Conde, J.; Rodriguez-Lopez, O.; Voit, W.E. Effect of annealing atmosphere on IGZO thin film transistors on a deformable softening polymer substrate. *Semicond. Sci. Technol.* **2018**, *33*, 095001. [CrossRef]
17. Huang, Y.; Yang, P.; Huang, H.; Wang, S.; Cheng, H. Effect of the annealing ambient on the electrical characteristics of the amorphous InGaZnO thin film transistors. *J. Nanosci. Nanotechnol.* **2012**, *12*, 5625–5630. [CrossRef] [PubMed]
18. Wu, J.; Chen, Y.; Zhou, D.; Hu, Z.; Xie, H.; Dong, C. Sputtered oxides used for passivation layers of amorphous InGaZnO thin film transistors. *Mater. Sci. Semicond. Process.* **2015**, *29*, 277–282. [CrossRef]
19. Nguyen, C.; Trinh, T.; Dao, V.; Raja, J.; Jang, K.; Le, T.; Iftiquar, S.; Yi, J. Bias-stress-induced threshold voltage shift dependence of negative charge trapping in the amorphous indium tin zinc oxide thin-film transistors. *Semicond. Sci. Technol.* **2013**, *28*, 105014. [CrossRef]
20. Xie, H.; Xu, J.; Liu, G.; Zhang, L.; Dong, C. Development and analysis of nitrogen-doped amorphous InGaZnO thin film transistors. *Mater. Sci. Semicond. Process.* **2017**, *64*, 1–5. [CrossRef]
21. Zhou, X.; Han, D.; Dong, J.; Li, H.; Yi, Z.; Xing, Z.; Wang, Y. The effects of post annealing process on the electrical performance and stability of Al-Zn-O thin-film transistors. *IEEE Electron Device Lett.* **2020**, *41*, 569–572. [CrossRef]
22. Chun, M.; Um, J.; Park, M.; Chowdhury, M.; Jang, J. Effect of top gate potential on bias-stress for dual gate amorphous indium-gallium-zinc-oxide thin film transistor. *AIP Adv.* **2016**, *6*, 075217. [CrossRef]
23. Nomura, K.; Kamiya, T.; Hosono, H. Effects of diffusion of hydrogen and oxygen on electrical properties of amorphous oxide semiconductor, In-Ga-Zn-O. *ECS J. Solid State Sci. Technol.* **2012**, *2*, 5–8. [CrossRef]
24. Yang, J.; Cao, D.; Lin, D.; Liu, F. Effect of annealing ambient gases on the bias stability of amorphous SnSiO thin-film transistors. *Semicond. Sci. Technol.* **2020**, *35*, 115003. [CrossRef]
25. Su, J.; Yang, H.; Ma, Y.; Li, R.; Jia, L.; Liu, D.; Zhang, X. Annealing atmosphere-dependent electrical characteristics and bias stability of N-doped InZnSnO thin film transistors. *Mater. Sci. Semicond. Process.* **2020**, *113*, 105040. [CrossRef]
26. Chen, Y.; Chang, T.; Li, H.; Chen, S.; Chung, W.; Chen, Y.; Tai, Y.; Tseng, T.; Huang, F. Surface states related the bias stability of amorphous In–Ga–Zn–O thin film transistors under different ambient gasses. *Thin Solid Films* **2011**, *520*, 1432–1436. [CrossRef]
27. Dong, C.; Xu, J.; Zhou, Y.; Zhang, Y.; Xie, H. Light-illumination stability of amorphous InGaZnO thin film transistors in oxygen and moisture ambience. *Solid-State Electron.* **2019**, *153*, 74–78. [CrossRef]
28. Pearton, S.; Cho, H.; LaRoche, J.; Ren, F.; Wilson, R.; Lee, J. Oxygen diffusion into SiO2-capped GaN during annealing. *Appl. Phys. Lett.* **1999**, *75*, 2939–2941. [CrossRef]
29. Nguyen, T.; Renault, O.; Aventurier, B.; Rodriguez, G.; Barnes, J.; Templier, F. Analysis of IGZO thin-film transistors by XPS and relation with electrical characteristics. *J. Disp. Technol.* **2013**, *9*, 770–774. [CrossRef]
30. Chen, W.; Lo, S.; Kao, S.; Zan, H.; Tsai, C.; Lin, J.; Fang, C. Oxygen-dependent instability and annealing/passivation effects in amorphous In–Ga–Zn–O thin-film transistors. *IEEE Electron Device Lett.* **2011**, *32*, 1552–1554. [CrossRef]
31. Qu, M.; Chang, C.; Meng, T.; Zhang, Q.; Liu, P.; Shieh, H. Stability study of indium tungsten oxide thin-film transistors annealed under various ambient conditions. *Phys. Status Solidi* **2016**, *214*, 1600465. [CrossRef]
32. Nomura, K.; Kamiya, T.; Hirano, M.; Hosono, H. Origins of threshold voltage shifts in room-temperature deposited and annealed a-In–Ga–Zn–O thin-film transistors. *Appl. Phys. Lett.* **2009**, *95*, 013502. [CrossRef]
33. Zhou, Y.; Dong, C. Influence of passivation layers on positive gate bias-stress stability of amorphous InGaZnO thin-film transistors. *Micromachines* **2018**, *9*, 603. [CrossRef]

34. Lee, H.; Cho, W. Effects of vacuum rapid thermal annealing on the electrical characteristics of amorphous indium gallium zinc oxide thin films. *AIP Adv.* **2018**, *8*, 015007. [CrossRef]
35. Park, Y.; Um, J.; Mativenga, M.; Jang, J. Thermal stability improvement of back channel etched a-IGZO TFTs by using fluorinated organic passivation. *ECS J. Solid State Sci. Technol.* **2018**, *7*, 123–126. [CrossRef]
36. Kim, C.; Kim, E.; Lee, M.; Park, J.; Seol, M.; Bae, H.; Bang, T.; Jeon, S.; Jun, S.; Park, S.; et al. Electrothermal annealing (ETA) method to enhance the electrical performance of amorphous-oxide-semiconductor (AOS) thin-film transistors (TFTs). *ACS Appl. Mater. Interfaces* **2016**, *8*, 23820–23826. [CrossRef] [PubMed]

Article

N-Type Nanosheet FETs without Ground Plane Region for Process Simplification

Khwang-Sun Lee and Jun-Young Park *

School of Electronics Engineering, Chungbuk National University, Cheongju 28644, Korea; ksunlee@chungbuk.ac.kr

* Correspondence: junyoung@cbnu.ac.kr

Abstract: This paper proposes a simplified fabrication processing for nanosheet Field-Effect Transistors (FETs) part of beyond-3-nm node technology. Formation of the ground plane (GP) region can be replaced by an epitaxial grown doped ultra-thin (DUT) layer on the starting wafer prior to $Si_x/SiGe_{1-x}$ stack formation. The proposed process flow can be performed in-situ, and does not require changing chambers or a high temperature annealing process. In short, conventional processes such as ion implantation and subsequent thermal annealing, which have been utilized for the GP region, can be replaced without degrading device performance.

Keywords: band-to-band tunneling; epitaxial growth; ground plane region; gate-all-around field-effect-transistors (GAA FETs); nanosheet FETs (NS FETs); parasitic channel leakage; punch-through

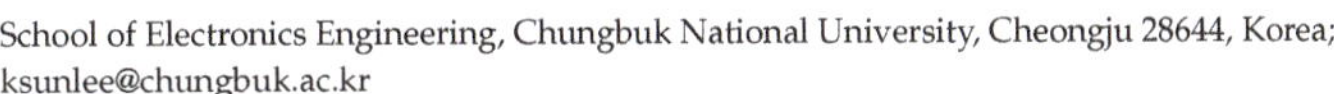

Citation: Lee, K.-S.; Park, J.-Y. N-Type Nanosheet FETs without Ground Plane Region for Process Simplification. *Micromachines* **2022**, *13*, 432. https://doi.org/10.3390/mi13030432

Academic Editor: Chengyuan Dong

Received: 11 February 2022
Accepted: 10 March 2022
Published: 11 March 2022

1. Introduction

To suppress short-channel effects (SCEs), which are an important concern with aggressive device scaling, semiconductor devices have evolved from 2-dimensional (2-D) structures to 3-dimensional (3-D) architectures. This has further benefits. For example, due to their superior gate controllability, FinFETs have lower subthreshold swing (*SS*) and off-state leakage (I_{OFF}) than planar devices [1–3]. However, as device scaling approaches extreme levels, it is becoming difficult to control SCEs using FinFETs. As a solution, nanosheet FETs (NS FETs), which have multiple channels with a gate-all-around (GAA) backbone structure, have been introduced as beyond-FinFETs [4–6]. However, even though NS FETs have shown better suppression of SCEs than FinFETs, as well as better output performance, there are still many difficulties related to mass production. Producers such as Samsung Electronics Inc. and TSMC Inc. are planning the mass production of NS FETs in 2022 and 2023, respectively. However, it is unknown whether the yield will be sufficient. The reason for this lies in difficulties in the fabrication processing of the NS FETs. For example, surface roughness scattering stemming from germanium diffusion among the nanosheets, striction stemming from adhesion between nanosheets [6], residue formation (e.g., TiN or Si_3N_4) due to uncontrollable wet etching, unwanted void formation during metal gate filling [7], etc., are very challenging problems in current fabrication processing.

In addition, with respect to source/drain (S/D) modules, the contact depth as well as inner spacer thickness needs to be optimized [8,9]. It should be noted in particular that, unlike bulk FinFETs, which have already been mass produced, NS FETs have a parasitic channel underneath the first floor nanosheet [10–13]. Hence, even though NS FETs have multiple channels that are completely surrounded by metal gates (i.e., GAA), increased I_{OFF} stemming from the parasitic channel is inevitable. V. Jegadheesan et al., have suggested that applying a ground plane (GP) region can effectively minimize the I_{OFF} in the parasitic channel [14]. However, as we have mentioned above, the fabrication process flow of NS FETs is very sensitive. For example, ion implantation and rapid thermal annealing (RTA) for GP region formation are associated with an unwanted non-uniform doping profile [15,16].

In this context, it would be better, in terms of device variability and yield, if the conventional processes could be replaced with an alternative process. However, recently, research papers covering the fabrication process of NS FETs have been modest in number.

In this letter, we propose a fabrication process flow for NS FETs. The proposed process flow does not require ion implantation or additional thermal treatment, which are conventionally performed to form the GP region. Alternatively, a doped ultra-thin (DUT) layer is epitaxially grown on the starting wafer in-situ before $Si_x/SiGe_{1-x}$ stack formation. This achieves process simplification, improved variability, and improved yield during the fabrication of NS FETs.

2. Materials and Method

The proposed fabrication process flow is summarized in Figure 1. In the conventional NS FETs fabrication process, ion implantation and thermal annealing are performed in dashed step two for GP region formation. However, the process can be replaced by Si epitaxial growth for DUT layer formation, as described in the bolded step two. The DUT layer is doped with a p–type dopant (i.e., boron) to minimize the parasitic channel and punch-through underneath the first floor nanosheet. The thickness of the DUT is similar to the thickness of silicon and Si_xGe_{1-x} layers, which are epitaxially grown as nanosheets and sacrificial layers, respectively. Hence, the formation of the DUT layer is not problematic under current processing technology. Above all, it should be noted that the processing from steps one to three in Figure 1 can be performed in-situ without changing chambers.

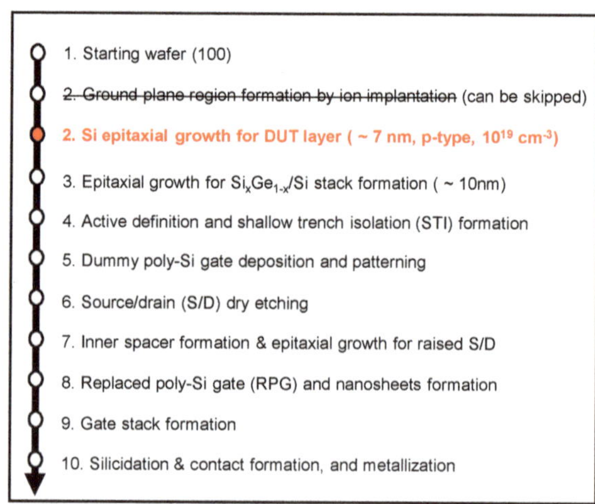

Figure 1. Fabrication process flow of proposed NS FETs including epitaxially grown DUT layer prior to formation of the Si_xGe_{1-x}/Si stack. The bolded step two can be alternatively added instead of the under-lined ion implantation and annealing process.

A 3-D simulator (Synopsys Sentaurus, Mountain View, CA, USA) was utilized to simulate fabrication processing and device characteristics. The drift-diffusion carrier transport equation was combined with the Poisson equation, and the density-gradient model was considered to reflect the quantum confinement effect of the nanosheet channels [17–20]. The Slotboom model was included for doping-dependent bandgap narrowing in the overall region [21]. Thin layer models such as inversion and accumulation layer mobility model (IALMob) were included to reflect impurity and phonon scattering [22]. In addition, the Shockley-Read-Hall (SRH) and non-local band-to-band tunneling (BTBT) recombination models were included to reflect gate-induced drain leakage (GIDL) during the simulations [22].

Figure 2 shows the backbone structure of the NS FETs used in the process simulation. Channel thickness (T_{CH}) and width (W_{NS}) were 5 nm and 45 nm, respectively. To elaborate, the dielectric constant of the HfO_2 gate dielectric and effective-oxide-thickness (EOT) were assumed to be 25 and 0.7 nm, respectively. In terms of doping concentration, epitaxially grown S/D regions were doped with arsenic at 3×10^{20} cm^{-3} [14,23]. The three nanosheet channels and silicon substrate (N_{Sub}) were lightly doped with boron at 1×10^{17} cm^{-3} and 1×10^{16} cm^{-3}, respectively. The doping concentration of the DUT layer was 1×10^{19} cm^{-3} of boron. The work-function of the titanium nitride for the metal gate was 4.5 eV. Detailed device parameters used for the simulations are summarized in Table 1.

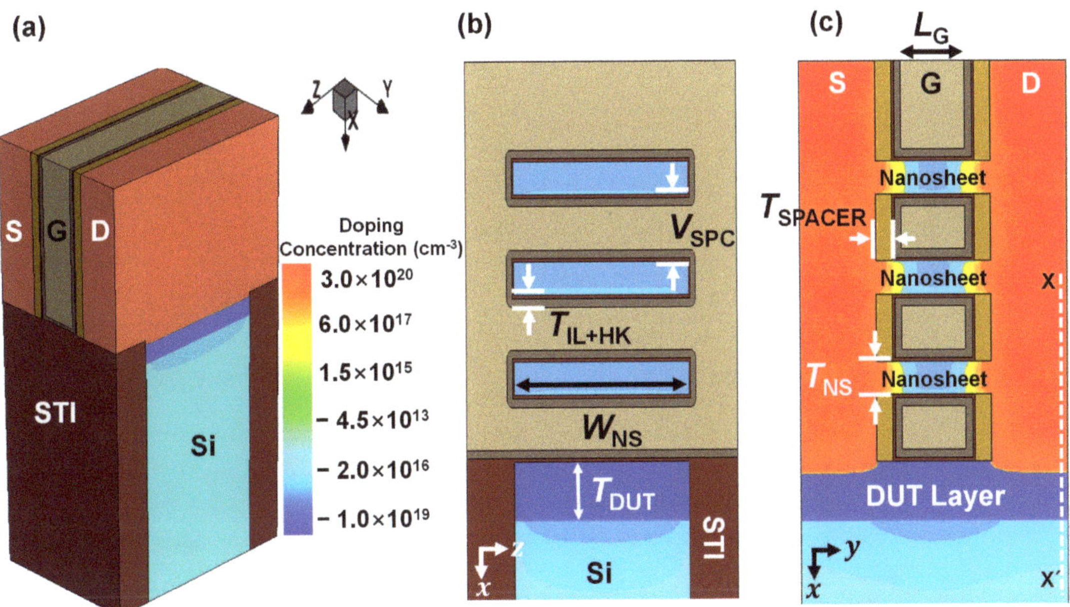

Figure 2. (**a**) Simulated NS FET device structure including the epitaxially–grown DUT layer on the starting wafer. Cross–sectional view of the proposed NS FET with a cut along the (**b**) gate and (**c**) channel directions, respectively.

Table 1. Dimensions and parameters used for the TCAD simulations.

Parameter	Value
Gate Length, L_G	12 nm
Nanosheet Width, W_{NS}	45 nm
Inner Spacer Thickness, T_{SPACER}	3 nm
Nanosheet-to-Nanosheet Vertical Space, V_{SPC}	10 nm
Nanosheet Thickness, T_{NS}	5 nm
Doped Ultra-Thin (DUT) Layer Thickness, T_{DUT}	5–100 nm
Doping Concentration of DUT Layer (N_{DUT})	10^{19} cm^{-3}
Inter Layer SiO_2 Thickness, T_{IL}	0.5 nm
High-k Gate Dielectric Thickness, T_{HK}	1.28 nm
Contacted Poly-Si Pitch (CPP)	44 nm

Then, the simulated I_D-V_G was carefully calibrated based on fabricated devices (i.e., such as doping concentration, structure, and gate work function) reported in [6], as shown in Figure 3a. Threshold voltage (V_{TH}) was extracted using a constant current method at I_D of 100 nA. Multiple V_{TH} characteristic can be visible when series resistance of NS FET is high due to defect existence or low doping concentration of S/D regions, as well as excessively long V_{SPC}. However, there was no observable multiple value of V_{TH} observed in the subthreshold region. Hence such concerns were not problematic, as shown in Figure 3b.

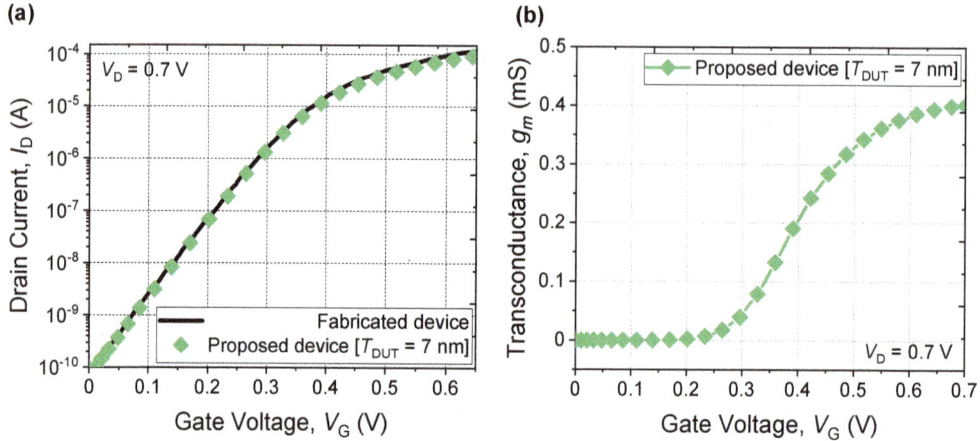

Figure 3. (**a**) Calibration of I_D-V_G curve with the fabricated device in Reference [6]. (**b**) Transconductance of (**a**).

3. Results and Discussion

Figure 4a shows the simulated doping profile during the off-state when the GP region as well as the DUT layer were not included. An unwanted parasitic channel and punch-through were formed underneath the first floor nanosheet. These concerns led to increased I_{OFF} during the off-state (Figure 4b).

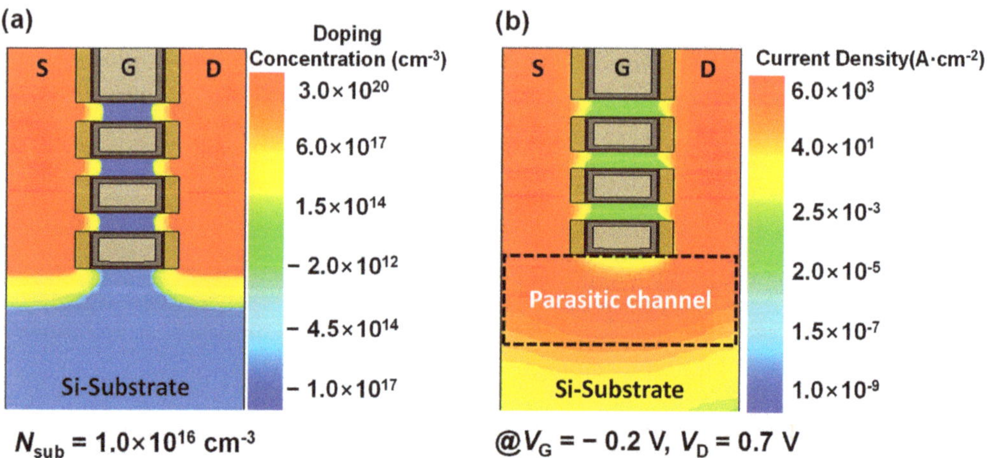

Figure 4. Simulated (**a**) doping profile distribution and (**b**) current density of a NS FET without a DUT layer during off-state.

Figure 5a shows the simulated transfer characteristics of the NS FETs with various thicknesses of DUT layers (T_{DUT}) from 0 nm to 7 nm. As the thickness of the DUT layer increases, $I_{OFF,}$ as well as SS, improve. The I_{OFF} extracted at $V_G = -0.2$ V was 21.8 nA with a 0 nm DUT layer but improved to 0.98 pA with a 7 nm DUT layer. Figure 5b shows the current density profile of the NS FETs in Figure 5a extracted at the off-state with $V_D = 0.7$ V, $V_G = -0.2$ V. As the T_{DUT} increases, I_{OFF} flowing through the parasitic channel (Y–Y′ direction) can be suppressed, aided by the increased energy barrier height (Φ_{bi} of the parasitic channel), as shown in Figure 5c. In addition, punch-through can be suppressed by applying a DUT layer.

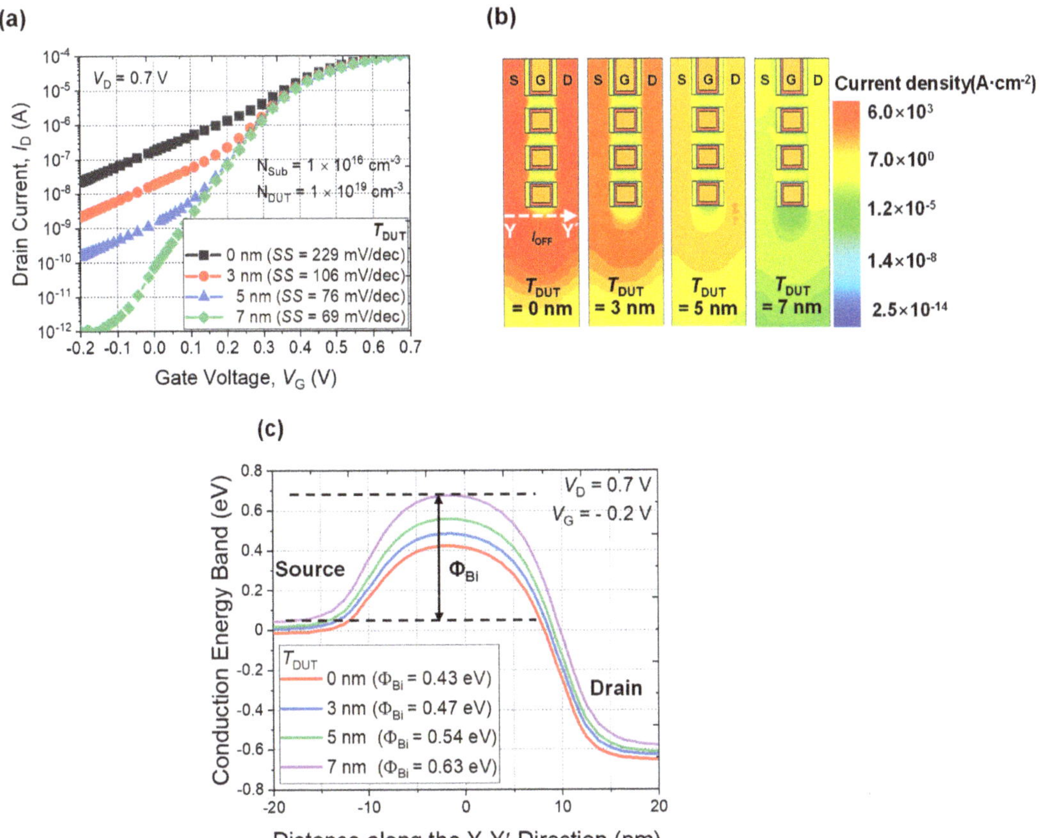

Figure 5. (**a**) Simulated I_D-V_G characteristic of NS FETs with various thicknesses of DUT layers without GP implantation and subsequent annealing. (**b**) Simulated parasitic current distribution profiles with various thicknesses of DUT layers. (**c**) Energy band diagram of parasitic channel along the Y–Y′ direction in (**b**). Conduction energy band height increases as T_{DUT} increases.

However, when T_{DUT} is thicker than 7 nm, the I_{OFF} increases (Figure 6a). Considering the I_{OFF} value in the range of V_G of –0.2 V to 0 V are identical to the drain–to–substrate leakage current, the source of the I_{OFF} increase is not the parasitic channel, but rather the BTBT between the drain and the DUT layer. In other words, when T_{DUT} is thicker than 7 nm, the DUT forms a p-n diode between the drain and the DUT layer, as shown in Figure 6b. During the off-state, the reverse biased p-n diode triggers BTBT of electrons, and increases the I_{OFF}. Figure 6c shows the simulated rate of electron generation between the drain and the substrate by the BTBT. As the T_{DUT} increases, there is a noticeable increase in electron generation by the BTBT. As a result, it can be concluded that a 7 nm thickness DUT layer is optimal to avoid unwanted increasing I_{OFF}.

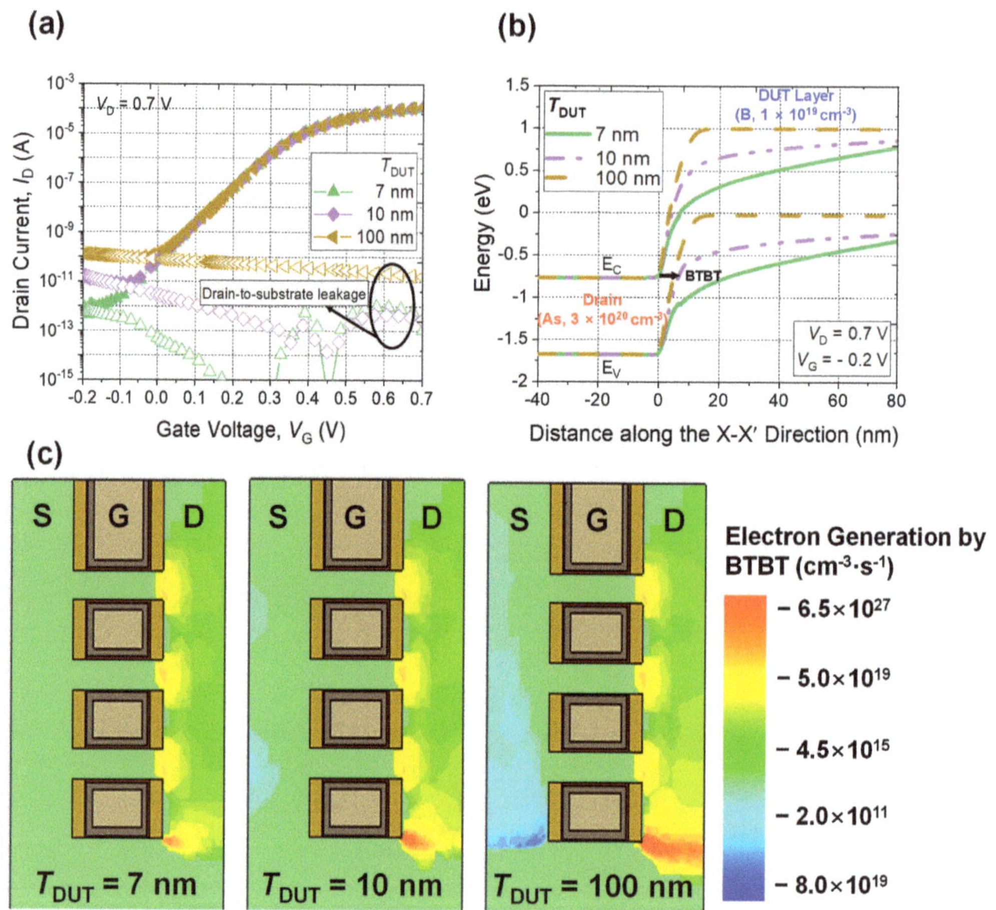

Figure 6. (**a**) Simulated I_D-V_G characteristic and (**b**) extracted energy band diagram of NS FETs (X–X′ direction in Figure 2c) with various thicknesses of DUT layers greater than 7 nm. (**c**) Electron generation rates with various T_{DUT}.

Figure 7a shows a simulated I_D-V_G curve for various doping concentrations (N_{DUT}) of the DUT layer. When the N_{DUT} is a low doping concentration of 1×10^{18} cm^{-3}, the I_{OFF} cannot be controlled due to increased *SS*. However, as the N_{DUT} increases, leakage current through the parasitic channel can be suppressed by the increased built-in potential (Φ_{bi}) from 0.39 eV to 0.59 eV, as shown in Figure 7b. In addition, suppression of punch-through is possible with increased substrate doping concentration.

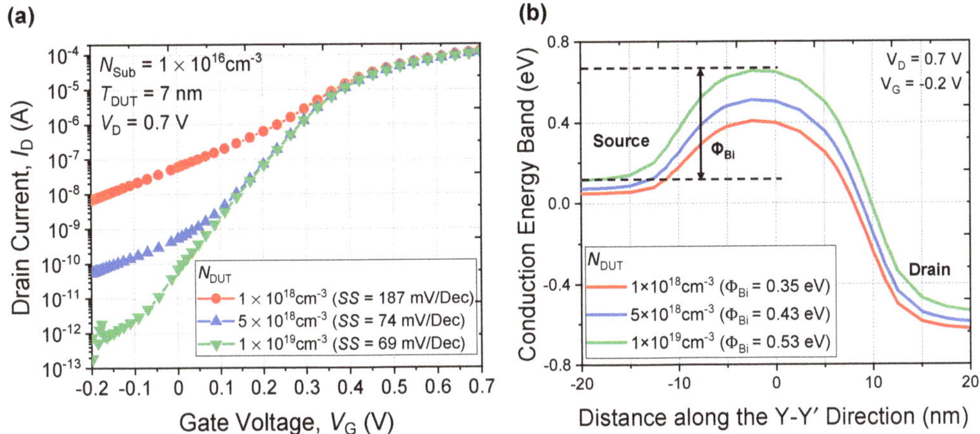

Figure 7. (**a**) I_D-V_G characteristic with different N_{DUT} layer. (**b**) Energy band diagram of the parasitic channel.

Figure 8 shows extracted I_D-V_G curve at V_D = 50 mV and 0.7 V, as well as I_D-V_D curve. The drain-induced barrier lowering (DIBL) of the proposed NS FET is 35 mV, which a low value compared with the value of 33 nm of the planar FET [19] or 15 nm of FinFET [24], as summarized in Table 2. In other words, the superior gate controllability of the NS FET does not degrade even when a DUT layer is applied.

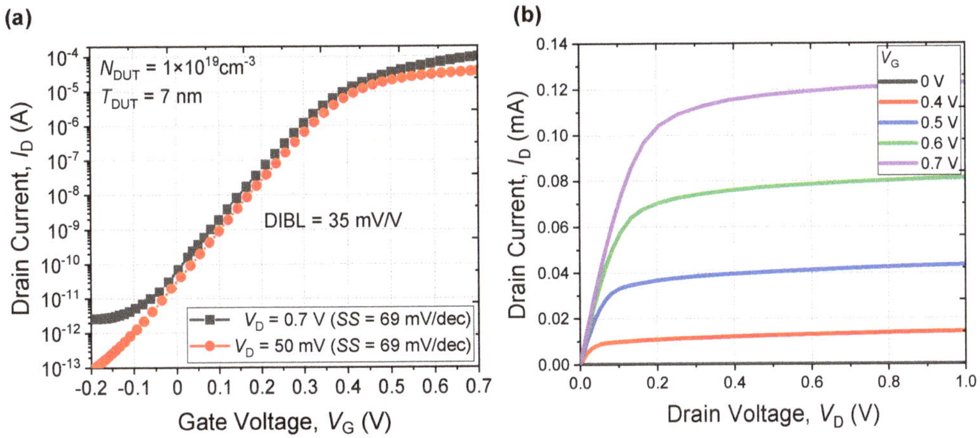

Figure 8. (**a**) Extracted I_D-V_G and (**b**) I_D–V_D characteristics of NS FET with DUT layer.

Table 2. Comparison of DIBL and *SS* characteristics with various devices.

	Planar FET [19]	FinFET [24]	NS FET [This Work]
DIBL (mV/V) (V_D = 50 mV and 0.7 V)	208	89	35
SS (mV/dec)	-	72	69

Figure 9 introduces various approaches to suppress the leakage current from the parasitic channel without the formation of a GP region [14,25]. A full bottom dielectric

(FBD) can eliminate the parasitic channel perfectly (Figure 9a). However, fabrication of the FBD inevitably requires a starting wafer composed of silicon-on-insulator (SOI) which is vulnerable to self-heating as well as expensive wafer cost. In this context, a steep-retrograde (SSR) region which contains a deep and heavily doped layer is preferred. However, forming an abrupt doping profile for the SSR is impossible using ion implantation and spike annealing. Even though epitaxial growth can be utilized, at least one more epitaxial growth step is required compared to the DUT case. Figure 9d shows the simulated electrical characteristics of the FBD, SSR, and DUT structures, respectively. Even though the DUT layer has a simpler fabrication process than the others, device characteristics in terms of V_{TH}, SS, and I_{ON} are not remarkable. Exact device parameters are summarized in Table 3.

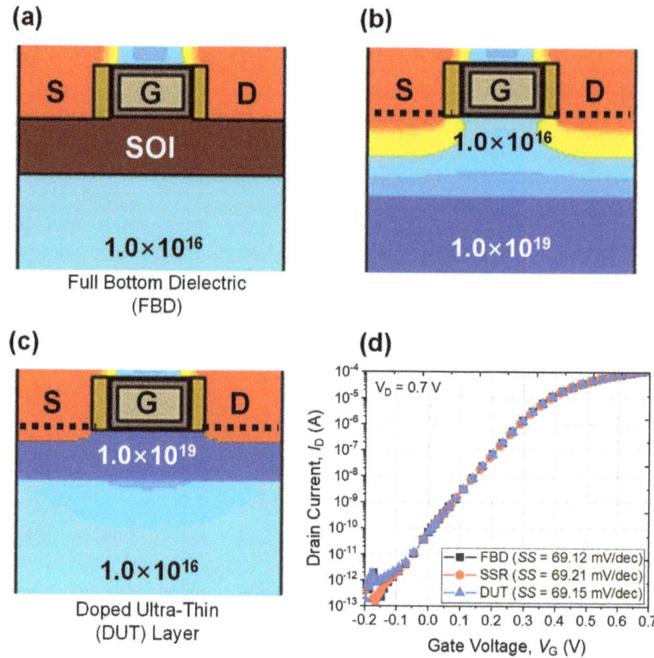

Figure 9. Various device structures without a GP region. NS FET with (**a**) full bottom dielectric (FBD) [25], (**b**) super steep-retrograde (SSR) region [14], and (**c**) the proposed DUT layer. (**d**) Simulated I_D-V_G characteristics of the respective device structures.

Table 3. Comparison of NS FETs based on FBD, SSR, and DUT structures.

	FBD	SSR	DUT
V_{TH} (0.7 V/50 mV) (mV)	212/229	212/233	216/239
SS (mV/dec)	69	69	69
I_{ON} (mA) at $V_G = 0.7$ V, $V_D = 0.7$ V	0.117	0.117	0.102
I_{OFF} (pA) at $V_G = 0$ V, $V_D = 0.7$ V	71.5	71.0	65.3
DIBL (mV/V) ($V_D = 50$ mV and 0.7 V)	32	32	35

4. Conclusions

For better process simplification, the fabrication process for nanosheet FETs was newly suggested based on 3-D simulation. The doped ultra-thin (DUT) layer can be epitaxially grown in-situ on the starting wafer. Conventional ground plane (GP) doping implantation as well as annealing process can be excluded while forming the DUT layer. The thickness of the DUT layer has been optimized to suppress parasitic channel leakage, punch-through, and band-to-band tunneling (BTBT). The NS FET with the DUT layer showed comparable performance, but has a simpler fabrication process compared with other NS FETs, including full bottom dielectric (FBD) or steep-retrograde region (SSR).

Author Contributions: For J.-Y.P. conceived this project and designed all of the experiments. K.-S.L. conducted all of the simulations and wrote this paper. All authors have read and agreed to the published version of the manuscript.

Funding: This work was supported by the National Research Foundation of Korea (No. 2020M3H2A 1076786 and 2021R1F1A1049456). The EDA tool was supported by the IDEC.

Data Availability Statement: Not applicable.

Conflicts of Interest: The authors declare no conflict of interest.

References

1. Chang, L.; Choi, Y.-K.; Ha, D.; Ranade, P.; Xiong, S.; Bokor, J.; Hu, C.; King, T.-J. Extremely scaled silicon nano-CMOS device. *Proc. IEEE* **2003**, *91*, 1860–1873. [CrossRef]
2. Razavieh, A.; Zeitzoff, P.; Nowak, E.J. Challenges and Limitations of CMOS Scaling for FinFET and beyond Architectures. *IEEE Trans. Nanotechnol.* **2019**, *18*, 999–1004. [CrossRef]
3. Hisamoto, D.; Lee, W.; Kedzierski, J.; Takeuchi, H.; Asano, K.; Kuo, C.; Anderson, E.; King, T.J.; Bokor, J.; Hu, C. FinFET—A self-aligned double-gate MOSFET scalable to 20 nm. *IEEE Trans. Electron Devices* **2000**, *12*, 2320–2325.
4. Lee, S.Y.; Kim, S.M.; Yoon, E.J.; Oh, C.W.; Chung, I.; Park, D.; Kim, K. A novel multi-bridge-channel MOSFET (MBCFET): Fabrication technologies and characteristics. *IEEE Trans. Nanotechnol.* **2004**, *2*, 253–257.
5. Bae, G.; Bae, D.-I.; Kang, M.; Hwang, S.M.; Kim, S.S.; Seo, B.; Kwon, T.Y.; Lee, T.J.; Moon, C.; Choi, Y.M.; et al. 3nm GAA Technology featuring Multi-Bridge-Channel FET for Low Power and High Performance Applications. In Proceedings of the 2018 IEEE International Electron Devices Meeting (IEDM), San Francisco, CA, USA, 1–5 December 2018; IEEE: San Francisco, CA, USA; pp. 28.7.1–28.7.4.
6. Loubet, N.; Hook, T.; Montanini, P.; Yeung, C.; Kanakasabapathy, S.; Guillom, M.; Yamashita, T.; Zhang, J.; Miao, X.; Wang, J.; et al. Stacked nanosheet gate-all-around transistor to enable scaling beyond FinFET. In Proceedings of the 2017 IEEE Symposium on VLSI Technology, Kyoto, Japan, 5–8 June 2017; IEEE: Kyoto, Japan; pp. T230–T231.
7. Witters, L.; Veloso, A.; Ferain, I.; Demand, M.; Collaert, N.; Son, N.J.; Adelmann, C.; Meersschaut, J.; Vos, R.; Rohr, E.; et al. Multiple-Vt FinFET devices through La_2O_3 dielectric capping. In Proceedings of the IEEE International SOI Conference, New Paltz, NY, USA, 6–9 October 2008; IEEE: New Paltz, NY, USA; pp. 121–122.
8. Li, J.; Li, Y.; Zhou, N.; Xiong, W.; Wang, G. Study of Silicon Nitride Inner Spacer Formation in Process of Gate-all-around Nano-transistors. *Nanomaterials* **2020**, *10*, 793. [CrossRef] [PubMed]
9. Lee, K.-S.; Park, J.-Y. Inner Spacer Engineering to Improve Mechanical Stability in Channel-Release Process of Nanosheet FETs. *Electronics* **2021**, *10*, 1395. [CrossRef]
10. Ritzenthaler, R.; Mertens, H.; De Keersgieter, A.; Mitard, J.; Mocuta, D.; Horiguchi, N. Isolation of nanowires made on bulk wafers by ground plane doping. In Proceedings of the 2017 47th European Solid-State Device Research Conference (ESSDERC), Leuven, Belgium, 11–14 September 2017; IEEE: Leuven, Belgium; pp. 300–303.
11. Choi, Y.; Lee, K.; Kim, K.Y.; Kim, S.; Lee, J.; Lee, R.; Kim, H.-M.; Song, Y.S.; Kim, S.; Lee, J.-H.; et al. Simulation of the effect of parasitic channel height on characteristics of stacked gate-all-around nanosheet FET. *Solid State Electron.* **2020**, *164*, 107686. [CrossRef]
12. Hong, J.M.; Park, J.W.; Lee, J.W.; Ham, J.H.; Park, K.R.; Jeon, J.W. Alpha Particle Effect on Multi-Nanosheet Tunneling Field-Effect Transistor at 3-nm Technology Node. *Micromachines* **2019**, *10*, 847. [CrossRef] [PubMed]
13. Seon, Y.; Chang, J.; Yoo, C.; Jeon, J. Device and Circuit Exploration of Multi-Nanosheet Transistor for Sub-3 nm Technology Node. *Electronics* **2021**, *10*, 180. [CrossRef]
14. Jegadheesan, V.; Sivasankaran, K.; Konar, A. Optimized Substrate for Improved Performance of Stacked Nanosheet Field-Effect Transistor. *IEEE Trans. Electron Devices* **2020**, *67*, 4079–4084. [CrossRef]
15. Ang, K.-W.; Barnett, J.; Loh, W.-Y.; Huang, J.; Min, B.-G.; Hung, P.Y.; Ok, I.; Yum, J.H.; Bersuker, G.; Rodgers, M.; et al. 300 mm FinFET results utilizing conformal, damage free, ultra shallow junctions (Xj$\sim$5 nm) formed with molecular monolayer

doping technique. In Proceedings of the 2011 IEEE International Electron Devices Meeting (IEDM), Washington, DC, USA, 5–7 December 2011; IEEE: Waschington, DC, USA; Volume 1–35, p. 35.

16. Chen, W.C.; Lin, H.C.; Chang, Y.C.; Lin, C.D.; Huang, T.Y. In Situ Doped Source/Drain for Performance Enhancement of Double-Gated Poly-Si Nanowire Transistors. *IEEE Trans. Electron Devices* **2010**, *57*, 1608–1615. [CrossRef]
17. Ancona, M.G.; Iafrate, G.J. Quantum correction to the equation of state of an electron gas in a semiconductor. *Phys. Rev. B.* **1989**, *39*, 9536–9540. [CrossRef] [PubMed]
18. Huang, H.S.; Wang, W.L.; Wang, M.C.; Chao, Y.H.; Wang, S.J.; Chen, S.Y. I-V model of nano nMOSFETs incorporating drift and diffusion current. *Vacuum* **2018**, *155*, 76–82. [CrossRef]
19. Chao, S.-Y.; Huang, H.-S.; Huang, P.-R.; Lin, C.-Y.; Wang, M.-C. Channel Mobility Model of Nano-Node MOSFETs Incorporating Drain-and-Gate Electric Fields. *Crystals* **2022**, *12*, 295. [CrossRef]
20. Woltjer, R.; Tiemeijer, L.; Klaassen, D. An industrial view on compact modeling. *Solid-State Electron.* **2007**, *51*, 1572–1580. [CrossRef]
21. Slotboom, J.W.; De Graaff, H.C. Bandgap Narrowing in Silicon Bipolar Transistors. *IEEE Trans. Electron Devices* **1977**, *24*, 1123–1125. [CrossRef]
22. *Sentaurus Device User Guide, Version L-2016.03*; Synopsys Inc.: Mountain View, CA, USA, March 2016.
23. Yoon, J.-S.; Jeong, J.S.; Lee, S.H.; Lee, J.J.; Lee, S.G.; Lim, J.W.; Baek, R.H. DC Performance Variations by Grain Boundary in Source/Drain Epitaxy of sub-3-nm Nanosheet Field-Effect Transistors. *IEEE Access*, 2022; *in press*. [CrossRef]
24. Xie, R.; Montanini, P.; Akarvardar, K.; Tripathi, N.; Haran1, B.; Johnson, S.; Hook, T.; Hamieh, B.; Corliss, D.; Wang, J.; et al. A 7 nm FinFET technology featuring EUV patterning and dual strained high mobility channels. In Proceedings of the 2016 IEEE International Electron Devices Meeting (IEDM), San Francisco, CA, USA, 3–7 December 2016; IEEE: San Francisco, CA, USA; pp. 2–7.
25. Zhang, J.; Frougier, J.; Greene, A. Full Bottom Dielectric Isolation to Enable Stacked Nanosheet Transistor for Low Power and High Performance Applications. In Proceedings of the IEEE 2019 IEEE International Electron Devices Meeting (IEDM), San Francisco, CA, USA, 7–11 December 2019; IEEE: San Francisco, CA, USA.

 *micromachines*

MDPI

Article

Efficient Multi-Material Structured Thin Film Transfer to Elastomers for Stretchable Electronic Devices

Xiuping Ding [1] and Jose M. Moran-Mirabal [1,2,*]

1 Department of Chemistry & Chemical Biology, McMaster University, 1280 Main Street West, Hamilton, ON L8S 4M8, Canada; dingx12@mcmaster.ca
2 Brockhouse Institute for Materials Research, McMaster University, 1280 Main Street West, Hamilton, ON L8S 4M8, Canada
* Correspondence: mirabj@mcmaster.ca

Abstract: Stretchable electronic devices must conform to curved surfaces and display highly reproducible and predictable performance over a range of mechanical deformations. Mechanical resilience in stretchable devices arises from the inherent robustness and stretchability of each component, as well as from good adhesive contact between functional and structural components. In this work, we combine bench-top thin film structuring with solvent assisted lift-off transfer to produce flexible and stretchable multi-material thin film devices. Patterned wrinkled thin films made of gold (Au), silicon dioxide (SiO_2), or indium tin oxide (ITO) were produced through thermal shrinking of pre-stressed polystyrene (PS) substrates. The wrinkled films were then transferred from the PS to poly(dimethylsiloxane) (PDMS) substrates through covalent bonding and solvent-assisted dissolution of the PS. Using this approach, different materials and hybrid structures could be lifted off simultaneously from the PS, simplifying the fabrication of multi-material stretchable thin film devices. As proof-of-concept, we used this structuring and transfer method to fabricate flexible and stretchable thin film heaters. Their characterization at a variety of applied voltages and under cyclic tensile strain showed highly reproducible heating performance. We anticipate this fabrication method can aid in the development of flexible and stretchable electronic devices.

Keywords: flexible electronics; wrinkling; shape-memory polymer; lift-off; hybrid structure; multilayer conductive films; wearable electronics

Citation: Ding, X.; Moran-Mirabal, J.M. Efficient Multi-Material Structured Thin Film Transfer to Elastomers for Stretchable Electronic Devices. *Micromachines* **2022**, *13*, 334. https://doi.org/10.3390/mi13020334

Academic Editor: Chengyuan Dong

Received: 26 January 2022
Accepted: 19 February 2022
Published: 20 February 2022

1. Introduction

The implementation of stretchable electronics can enable new types of devices for a range of applications, like devices that can integrate with the human body for advanced therapeutic treatment, sensory skins for robotics, and wearable communication devices, which are impossible to achieve with conventional rigid electronics [1]. Flexible and stretchable components are key to the development of such devices for biomedicine, as has been shown in devices for intracardial and neural monitoring as well as human/machine interface integrated circuits [2]. Stretchable electronic devices should be mechanically robust to make it possible to conformally span curved surfaces and movable parts [3,4], and should be able to withstand large strains with no fracture or substantial degradation of their electrical properties [5]. For example, in a humanoid robot, areas covering the shoulders, elbows, and knees should be able to deform to 130–160% of their original dimensions [6]. Thin films have been used to implement stretchable electronic components because when the thickness of a thin film becomes 1/1000 of the desired radii of curvature, the tensile and compressive strains on the film during bending are small and the film becomes flexible and rollable [3,7].

Wrinkled structures, arising from the buckling mechanics of thin films, have been used to fabricate heterogeneous metal-elastomer stretchable electronic devices. The basis of this approach is that a supported thin film responds to an applied strain by buckling due to the stiffness mismatch between the rigid film and the compliant substrate. In such buckled materials, changes in amplitude and wavelength can occur reversibly when they are stretched or compressed [8]. For example, buckles with uniform 20–50 μm wavelengths have been created by thermally expanding a PDMS substrate, depositing a gold film by e-beam evaporation, and cooling down the system after gold deposition [9]. Similarly, stretchable and foldable silicon circuits have been fabricated by transferring the circuits from silicon-on-insulator wafer carrier substrates onto PDMS. In this approach, poly(methyl methacrylate) was spin coated as a sacrificial layer before the fabrication of the circuit arrays, and this layer was subsequently dissolved to release the ultrathin, flexible circuits during the transfer process [10].

The wrinkling of gold films by shrinking shape memory polymer substrates has been shown as an alternative cost-effective and simple method to make highly conductive stretchable electrodes. In this fabrication approach, thin gold films are deposited onto a shape-memory polymer substrate (e.g., polystyrene—PS) that is subsequently thermally shrunk [11], resulting in wrinkled thin films that can then be transferred from the rigid substrate to an elastomer like PDMS or Ecoflex [12,13]. In one example of this approach, the thin film transfer was done by dissolving a sacrificial photoresist layer under the gold film, manually lifting off the wrinkled film from the substrate, and depositing it onto partially cured PDMS. The advantage of a lift-off approach is that the photoresist layer can be quickly dissolved using acetone leaving behind minimal residues. A limitation of this method was the size of the transferred devices, as smaller, more fragile thin structures or thin films are difficult to transfer by hand without damage. A second method has been reported that avoids the use of a sacrificial layer and can faithfully transfer thin films patterned into complex shapes with small dimensions [12]. In this approach, the surface of a wrinkled gold film is treated with a solution of (3-mercaptopropyl)trimethoxysilane (MPTMS) to form a self-assembled monolayer that can covalently bond the film with a silicone elastomer mixture that is poured directly on top of it and cured. Through this method, highly stretchable wrinkled gold thin film electrodes were successfully fabricated. However, only a single layer gold film transfer was demonstrated, which limits the application of this method to more complex electronic devices and arrays, as many electronic devices are made in the form of multilayer composites.

In this work, we present two solvent-assisted lift-off methods for the transfer of various wrinkled thin film materials and multilayer structures from rigid PS substrates to elastomeric substrates. The process begins with the fabrication of structured thin films by shrinking thin films deposited onto pre-stressed PS sheets. The wrinkled films are then treated by plasma oxidation or through the formation of a self-assembled monolayer, which promotes the covalent binding of the thin films to a PDMS elastomer that is cast on top and cured. Finally, either the sacrificial layer is dissolved, or the PS substrate is swelled and partially dissolved to remove the film/PDMS from the rigid PS substrate. These methods are simple and effective ways of fabricating composite structures for stretchable electronic devices. They can reliably transfer complex patterns of materials, and different architectures with varied dimensions. The ability to transfer complex multi-material composites can open the door to applications that require stretchable electrodes, dielectric, and semiconductor components. It can also solve interconnect problems, which are one of the key challenges for wearable electronic devices and arrays [14].

As proof-of-concept of this fabrication approach, we have implemented Au and ITO/Au/ITO stretchable resistive heaters and characterized their response to applied voltage and strain. Both types of heaters show fast heating response, high robustness, and remarkable resilience through stretching and relaxation cycles. This type of stretchable heater can address the need for portable, comfortable, functional materials for physiotherapy, where thermal therapy is frequently used to alleviate joint pain caused by obesity, aging, or workplace injuries [15]. Thin film wearable heaters with high mechanical robustness and good Joule heating performance can be used to overcome the mechanical rigidity and weight of current heat packs or wraps used for this method of therapeutic care. We anticipate that the fabrication of stretchable electronic devices through solvent-assisted lift-off and transfer of composite wrinkled thin films can be a simple and cost-effective way to producing highly stable and stretchable wearable electronics.

2. Materials and Methods

2.1. Wrinkled Thin Film Fabrication

PS sheets (Graphix shrink film, Graphix, Maple Heights, OH, USA) were cut to desired sizes and cleaned by immersing them in isopropanol, ethanol, and 18.2 MΩ cm water (obtained from a Milli-Q Reference A+ Water Purification System, Millipore, Molsheim, France, subsequently referred to simply as "water") bath sequentially and washing them on an orbital shaker (MAXQ 2000, Thermo Fisher Scientific, Waltham, MA, USA), for 5 min for each solvent. The PS sheets (shape memory polymer) used in this work are commercially available and biaxially pre-stressed. As a result, the side length of a PS sheet can be shrunken to 40% of its original size by simply heating the substrate above the glass transition temperature. Mask stencils were created by cutting self-adhesive vinyl (FDC-4300, FDC graphic films, South Bend, IN, USA) with a Robo Pro CE5000-40-CRP cutter (Graphtec America Inc., Irvine, CA, USA) to create the desired patterns for different devices. The self-adhesive vinyl stencil was then applied onto the clean PS substrates. For the samples lifted off with a sacrificial layer, positive photoresist (Microposit S1805, Shipley, Marlborough, MA, USA) was spin coated onto the masked PS substrate at 7000 RPM for 30 s and baked at 90 °C for 3 min to remove the solvent in the photoresist. The thickness of the spin-coated photoresist was measured with a surface profilometer (Tencor Alpha Step 200, KLA Corp., Milpitas, CA, USA). Thin films were deposited onto masked PS substrates by sputtering using a Torr Compact Research Coater CRC-600 manual planar magnetron sputtering system (New Windsor, NY, USA). A 99.999% purity gold target (LTS Chemical Inc., Chestnut Ridge, NY, USA), SiO_2 target (Bayville Chemical Supply Company Inc., Deer Park, NY, USA), and ITO target (LTS Research Laboratories, Inc., Orangeburg, NY, USA) were used to deposit thin films of the respective materials. Gold films were sputtered using a DC (direct current) gun, and SiO_2 and ITO films were deposited by RF (radio frequency) gun. After peeling off the vinyl mask, the PS substrate with deposited films were heated in an oven at 130 °C for 5 min to induce biaxial shrinking and flatted on a silicon wafer by annealing at 160 °C for 10 min.

2.2. Lift-Off of Wrinkled Films from PS to PDMS

All samples with a gold film as the interface to be transferred to the elastomer were immersed in 5 mM (3-Mercaptopropyl) trimethoxysilane (MPTMS) (95% MPTMS, Sigma-Aldrich, MO, USA) aqueous solution for 1 h, rinsed with water and dried with nitrogen gas. Samples with SiO_2/ITO films interfacing the elastomer were treated using a Harrick High Power Plasma Cleaner (PDC, Harrick Plasma Inc., Ithaca, NY, USA) for 1 min at high power (30 W). At the same time, the base of PDMS and curing agent (Sylgard 184 silicone elastomer kit, Dow Corning Corporation, Corning, NY, USA) were fully mixed in a 10:1 mass ratio and degassed in a desiccator with a mechanical vacuum pump until air bubbles were removed. The degassed PDMS was cast onto the surface-treated structured films, then put in an oven at 60 °C for 4 h to be cured.

The cured PDMS was cut with a blade to expose the edges of PS substrate. Samples with and without photoresist were put into crystallization dishes with acetone and washed on an orbital shaker at 100 RPM. The photoresist was dissolved in acetone for approximately 1 h, at which point the PS substrate detached from the film/PDMS. For samples without a photoresist layer, the PS substrate was detached by swelling and softening PS in acetone for approximately 6 h, at which point the wrinkled thin films could be manually detached from the softened PS due to the strong chemical bonding between film and PDMS. However, a small amount of PS residue was visible on the surface of transferred films using this direct transfer method without photoresist. The PS residue was then removed by gently rinsing with toluene. The samples were then placed in the vacuum to extract excess solvent.

Although toluene is effective at removing PS residues, the PDMS also swelled significantly in this solvent and PS residues were observed via SEM on samples with short toluene rinsing times.

Hybrid films of Au/ITO/SiO_2 were made by the fabrication procedure described and transferred with no sacrificial layer.

2.3. Fabrication of Stretchable Thin Film Heaters

The stretchable heater is comprised of conductive pads made of 50 nm-thick Au films, and heating elements made of 50 nm-thick Au films or 5 nm-thick ITO/50 nm-thick Au/5 nm-thick ITO composite films. Single deposition was conducted for the Au heater; for the ITO/Au/ITO heaters, both ITO layers were deposited on the heating element area, and Au was deposited on the whole area. The thin films were deposited onto pre-stressed PS substrates, and subsequently heated in an oven at 130 °C for 1 min to shrink the sample and produce wrinkled films. The transfer process used was the same as the procedure described in Section 2.2.

2.4. Characterization of Stretchable Thin Film Heaters

To characterize the thermal and mechanical properties of the two types of thin film heaters, voltage was supplied by a source meter (2450 Source Meter, Keithley, Tektronix Inc., Solon, OH, USA). Thermal compound (ARCTIC MX-4, ARCTIC Ltd., Brunswick, Germany) was placed onto the heating element to absorb the heat and a thermocouple (T1 thermocouple, MS-6514 Thermometer, MASTECH-Group, Dongguan, China) was used to measure the temperature of the thermal compound. A home-built stretcher with the ability to control the stretch step and stretch distance was used to apply the external strain. The wires of the thermocouple were fixed onto a lab jack to ensure that the thermocouple would be reliably inserted into the thermal paste. The morphology of the thin films was characterized by scanning electron microscopy (SEM) imaging (JSM-7000F, JEOL Ltd., Tokyo, Japan), with a 10.0 kV acceleration voltage.

3. Results and Discussion

3.1. Thin Film Transfer with and without a Sacrificial Layer

Two solvent-assisted methods to transfer wrinkled thin films onto elastomeric substrates, one with a sacrificial layer and another without, were compared to evaluate the advantages and disadvantages of each approach. Figure 1 shows a schematic representation of the two processes tested to transfer wrinkled thin films (A and B) as well as a photo of sample patterned films transferred onto PDMS (C). Figure 1A shows the transfer process with a sacrificial layer of photoresist. To transfer the film with a sacrificial layer, a 400 nm-thick photoresist layer was spin-coated onto vinyl-masked PS prior to gold film deposition. The addition of the photoresist layer increased the wrinkle size compared to wrinkles formed on 20 nm-thick gold films deposited directly onto the PS substrate. This is because the wavelength of the wrinkles is proportional to the combined thickness of the stiff film and the sacrificial layer, and because the sacrificial layer can soften and flow during the shrinking process [11,16,17].

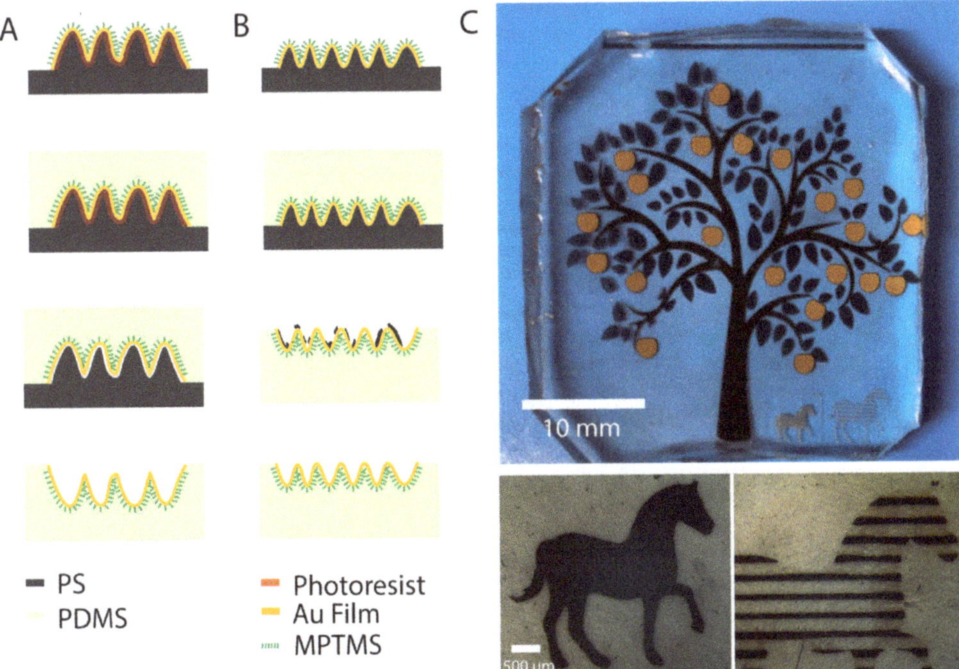

Figure 1. Schematic of two solvent-assisted processes to transfer wrinkled films from PS to PDMS. (**A**) Shrunken electrode with photoresist underneath as sacrificial layer. The surface of the wrinkled thin film was treated with MPTMS, and PDMS was cast onto the treated films and cured. The sacrificial layer was dissolved in acetone, completing the transfer of the wrinkled thin film to PDMS. (**B**) Shrunken electrode with no sacrificial layer on PS substrate. The surface of the wrinkled thin film was treated with MPTMS and PDMS was cast onto the treated films and cured. Incubation in acetone was used to swell and partially remove the PS, leaving some residues on the wrinkled surface. Washing with toluene was then used to remove the residues. (**C**) Thin films transferred onto PDMS with different compositions and thicknesses form an apple tree and horse scene: trunk—5 nm ITO/5 nm Au/5 nm ITO, leaves—1.3 nm ITO/1.3 nm Au/1.3 nm ITO, apples—50 nm Au, horses—20 nm SiO_2.

A strong bond between the thin films and the elastomer substrates is essential to impart the robustness and reliability needed for stretchable electronics. However, the bare gold film interacts very weakly with PDMS. To ensure strong adhesion between the wrinkled thin film and the elastomer, MPTMS was used to functionalize the gold surface. MPTMS is a molecular adhesion promoter that can form self-assembled monolayers and has two types of terminal groups with different functionalities. The three methoxy (-OCH_3) groups can covalently bind to the surface of PDMS and the thiol (-SH) can bond to the gold surface. Thus, when a PDMS base and crosslinker mixture was cast onto the functionalized surface of the shrunken electrode and cured at 60 °C for 4 h, the two materials became strongly bonded. Once the elastomer was cured, it was cut along the edge of the PS to expose the PS/PDMS interface. The PS with cured PDMS was then immersed in an acetone bath to dissolve the photoresist between the wrinkled Au film and the PS. The wrinkled films could be fully lifted off from the PS after only 1 h in acetone, while remaining adhered to the PDMS.

Figure 1B illustrates the transfer process that involves no sacrificial layer. In this example, a 20 nm-thick Au film was directly sputtered onto a PS substrate. In contrast to Figure 1A, the wavelength of the bare Au film (<1 μm) is much smaller than that of the

20 nm-thick Au film with a layer of photoresist (~4 μm). The wrinkled Au surface was also functionalized with MPTMS and PDMS was cast and cut using the same procedure as above. The cured sample of PS/Au/PDMS was put in an acetone bath for ~6 h. During this time, the PS substrate softened and swelled, and could be removed from the PDMS substrate. However, because of the softening of the PS, residue was left on the surface of the Au film. Since PS is more readily solubilized in toluene than acetone, as a last step, toluene was used to remove residue on the surface of the wrinkled film. Many toluene rinsing steps (~40) were needed to ensure that the rough wrinkled surface was devoid of any PS residue. Although toluene is useful to remove residual PS from the Au films, it causes swelling of PDMS by ~30% [18]. Swelled PDMS is, however, capable of shrinking back to its original dimensions by removing the solvent in a vacuum or through evaporation in an oven at 60 °C. The swelling introduced about 5% strain onto the wrinkled film and no cracks or delamination were observed on the transferred films.

To showcase the potential of transferring electronic devices with complexity by one of the methods described above, a pattern comprised of various materials in different shapes and layout was transferred. Figure 1C shows structured films transferred to PDMS made from various material layers and with different thicknesses: trunk of the apple tree—5 nm ITO/5 nm Au/5 nm ITO, leaves—1.3 nm ITO/1.3 nm Au/1.3 nm ITO, apples—50 nm Au, horses—20 nm SiO_2. In addition to different film thicknesses and compositions, the pattern transferred is complex and shows dimensions across different scales, from the base of the tree representing a couple millimeters and the stripes of the horse being ~120 μm in width. These films were transferred simultaneously and with no sacrificial layer, demonstrating the ability of this method to transfer different materials, hybrid structures, and multiscale complex patterns. This method is thus straightforward and high-fidelity, enabling the fabrication of stretchable thin film devices comprised of different materials and hybrid structures.

Both transfer methods have advantages and disadvantages based on the simplicity of the process and fidelity of the transferred patterns. The method using a sacrificial photoresist layer is simpler and more efficient, involving only the direct dissolution of photoresist with acetone, with no PS residue observed via SEM. However, the wavelength of the wrinkles is much larger compared to that of samples with similar Au film thickness transferred with the method devoid of photoresist. Figure 2A shows the surface topography of a 20 nm-thick gold wrinkled film transferred onto PDMS with a sacrificial layer. The increase in wrinkle size arises primarily from the greater combined film thickness (~400 nm for Au + photoresist versus 20 nm for Au only) as previously mentioned. The transfer process without sacrificial layer is longer and requires an extra toluene rinsing step to remove PS residue after the acetone bath. The residue, which can be seen in Figure 2B, fills the valleys on the wrinkled surface after removing the softened PS, but can be effectively eliminated with toluene (Figure 2C shows the same sample as Figure 2B after toluene rinsing). Therefore, the transfer method can be selected depending on the material requirements and the desired application. With the sacrificial layer, the transfer is fast and requires less post-transfer cleanup, but produces larger wrinkles whereas without the sacrificial layer, the transfer requires more time and an extra toluene rinsing stage, however the result is smaller wrinkled features. Each transfer method could prove useful depending on the requirements of roughness (wrinkle size), tolerance to PS residue and swelling, and compatibility with toluene for the stretchable electronic device.

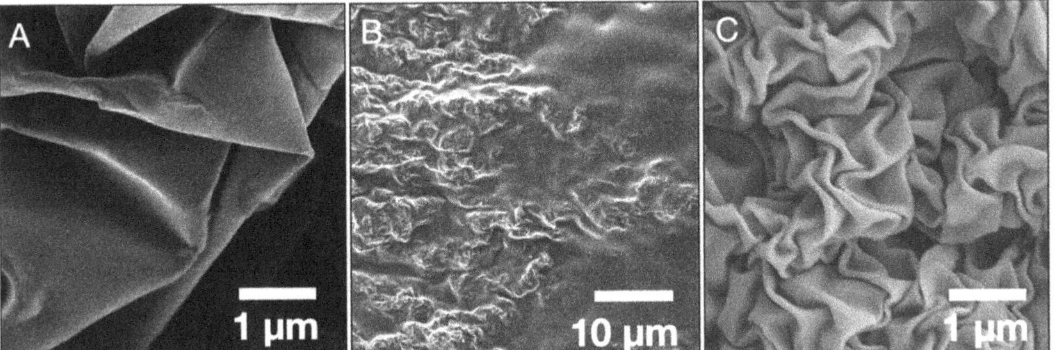

Figure 2. Scanning electron microscope images show the differences arising from of the two transferring methods. (**A**) Topography of a 20 nm-thick Au wrinkled film transferred onto PDMS using a sacrificial photoresist layer. Topography of a 20 nm-thick wrinkled gold film transferred without photoresist onto PDMS (**B**) with polystyrene residue after the acetone bath step and (**C**) after rinsing with toluene.

3.2. Multi-Material Thin Film Transfer

Most stretchable thin film devices or integrated circuits require the incorporation of multiple materials into layers with dimensions going from the macro to the microscale and functions ranging from conductor to dielectric. Therefore, it is essential that transfer methods can accommodate different materials and structures simultaneously. This would significantly simplify stretchable electronic device fabrication, achieve better adhesion between the different layers, and solve interconnection problems between different parts of the device, as well as prove useful for the fabrication of large area wearable electronic arrays.

To this end, we tested the ability of the direct transfer method to transfer composite layered thin films. SEM images of flat and wrinkled 5 nm ITO/5 nm Au/5 nm ITO film on PS substrates are shown in Figure 3A,C, respectively. The same films transferred onto PDMS are shown in Figure 3B,D, respectively. When measuring the resistance of transferred films, wrinkled films on PDMS showed the same values as on the rigid PS substrate. On the other hand, the transferred flat films (Figure 3B) were not electrically conductive due to extensive cracking. Both the morphology and electrical measurements suggest that the wrinkled structures make the thin film more tolerant to the strain applied during the transfer process. Figure 3C,D (ITO/Au/ITO films), along with Figure 3E,F (20 nm-thick SiO_2 on PS and PDMS), highlight the potential of this method for transferring multilayer, high modulus, and low fracture toughness wrinkled films from PS to PDMS. This will bring real benefit to the development of stretchable thin film electronic devices because most devices incorporate specific architectures composed of conductor, dielectric, and insulator materials. The one-time transfer will simplify the fabrication of stretchable devices and circuits, and significantly improve the bonding between different films. Moreover, the wrinkles increase the tolerance of modulus films to external strain during the fabrication, as shown by the comparison between Figure 3B,D.

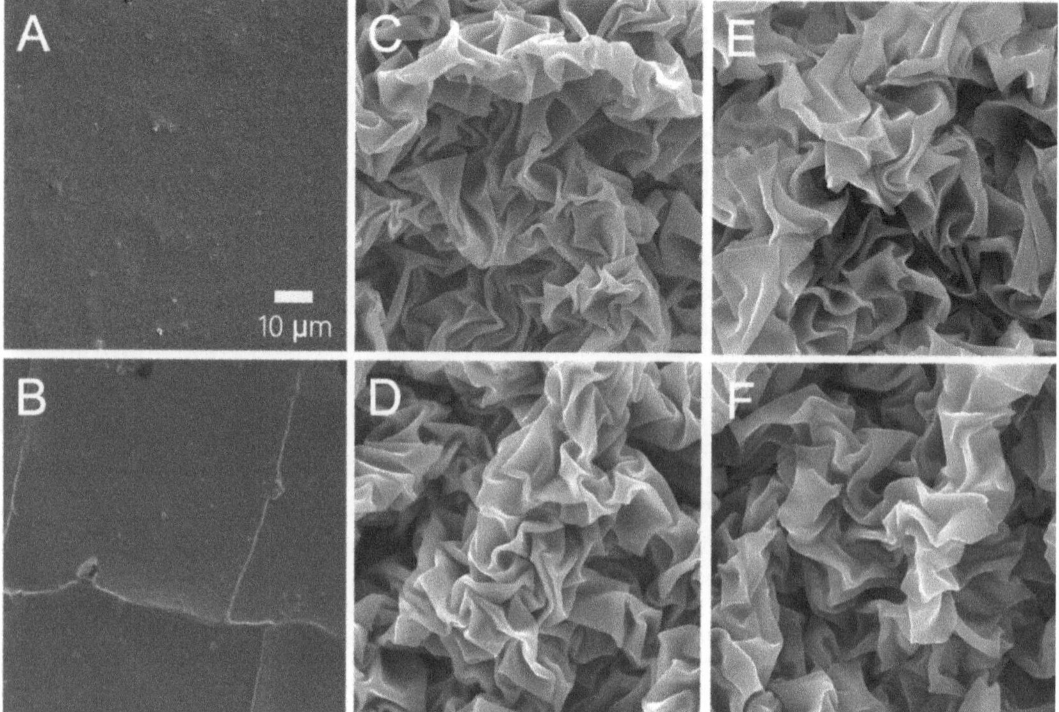

Figure 3. Surface topography of 5 nm ITO/5 nm Au/5 nm ITO multilayer films and 20 nm SiO_2 films on PS and PDMS substrates. (**A**) Flat ITO/Au/ITO film on PS. (**B**) Flat ITO/Au/ITO film transferred onto PDMS. (**C**) Wrinkled ITO/Au/ITO film on PS. (**D**) Wrinkled ITO/Au/ITO film on PDMS. (**E**) Wrinkled 20 nm SiO_2 on PS. (**F**) Wrinkled 20 nm SiO_2 on PDMS.

3.3. Stretchable Thin-Film Heaters

As proof-of-concept of the fabrication of stretchable devices through solvent-assisted transfer of structured thin films, stretchable thin film heaters were fabricated. Two types of heaters were built: heaters with a single layer of Au and heaters with multiple layers, composed of ITO/Au/ITO. The heaters were made containing a resistive heating pad (3×1.5 cm) and two small contact pads (0.5×0.5 cm). To characterize the mechanical and thermal properties of the stretchable heaters, thermal paste was applied on the surface of the resistive heating element, and the temperature of the thermal paste was monitored for various applied input voltages (Figure S1). The temperature vs. time profile for Au heaters under various applied voltages (0.2–1.7 V) over 420 s is shown in Figure 4A. At first the temperature of the thermal paste increases rapidly, especially during the first 60 s, then slows down and reaches a plateau (or stable temperature) within 100–200 s. Within the first 60 s, the temperature increases almost linearly, with heating rates for Au heaters of 0.25 °C/s and 0.70 °C/s under a bias of 1.0 V and 1.7 V, respectively. When the input voltage is below 1 V, the plateau temperature is reached faster than with higher applied voltages, as expected from the smaller change in temperature from room temperature. From the temperature-time profile it can be seen that the temperature of the heater could reach and maintain a stable temperature within minutes.

To investigate the stretchability of the structured Au film resistive elements, the heater was operated under various applied tensile strains at constant voltage and the plateau temperatures were measured. This was repeated for various applied voltages and the results are summarized in Figure 4B,C. The plot of temperature vs. strain (Figure 4B) shows mostly uniform horizontal lines for every applied voltage, indicating that Joule heating is stable when the heater is stretched. Figure 4C shows the same data, but now presented as a temperature vs. voltage plot. In this plot, it can be clearly observed that the heater presents a non-linear, quadratic (Figure S2), response to voltage. This functional dependence is maintained even at different applied strains, although higher voltages (higher temperatures) produce higher variability. This is expected as the heat can be quickly dissipated and small changes in the conductivity of the heating element can have a strong influence on the recorded temperature. It is worth noting that the temperatures recorded for strains at 5% show a slightly higher temperature compared to the temperature recorded under no applied strain. This can be explained by the way the stretchable heater is fixed in the stretching setup; when the thin film on the PDMS substrate was fixed on the stretcher and the screws that clamped the substrates were tightened, the PDMS substrate was squeezed, causing the film to undergo a small amount of compressive bending. The film was subsequently flattened during stretching, which means that the 0% strain state resulted in a slightly bent film and 5% strain state more accurately represented a relaxed "0% strain" state. To further check the temperature reproducibility, on-off cycles were applied to the heaters every 60 s under an applied voltage of 1.5 V (Figure 4D). The thin film heaters showed highly reproducible behavior, reliable performance, and fast heating response as indicated by the uniformity of the on/off response curve.

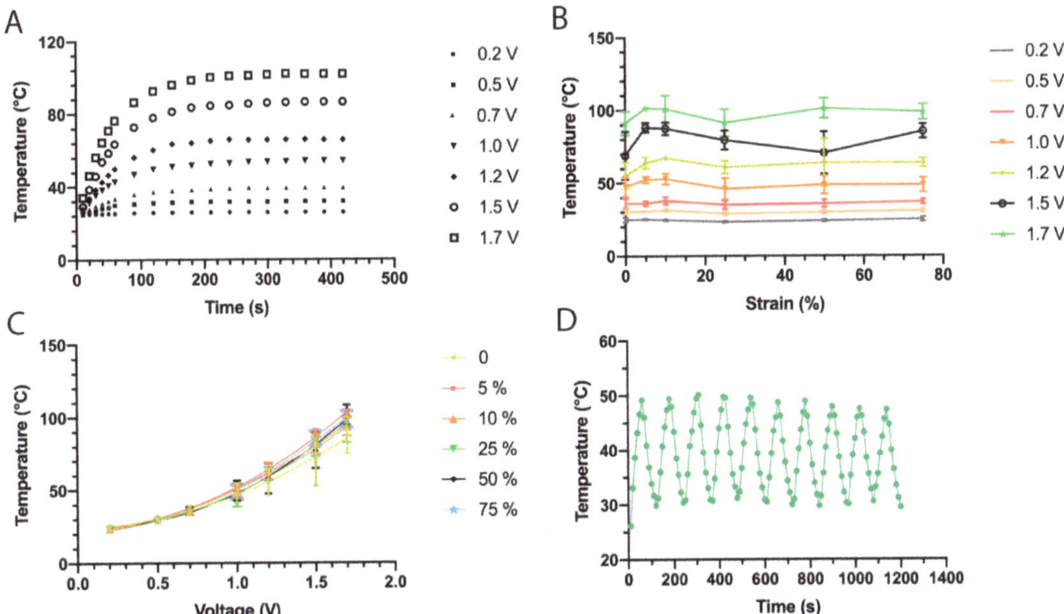

Figure 4. Characterization of stretchable Au heaters. (**A**) Temperature profiles under varying applied voltage (0.2–1.7 V). (**B**) The plateau temperature changes with various voltages but remains similar for any given voltage under different strains. (**C**) Changes in temperature vs. voltage for various applied strains (each series corresponds to an applied strain). (**D**) Cycling on/off response under an applied voltage of 1.5 V shows the reproducibility of the heating and cooling cycles.

To investigate the applicability of our solvent-assisted transfer fabrication of resistive heaters to composite ceramic/metal multilayered films, ITO/Au/ITO thin film heaters were fabricated and characterized. Although the heating element is comprised of ITO/Au/ITO multilayers, the contact pads were directly connected to the central Au heating layer. This fabrication step is very important to prevent the heater from burning because of a high contact resistance that could lead to overheating of the contact area. The composite thin film heaters were characterized in the same way as the Au heaters described above. Overall, similar thermal and mechanical characteristics were observed as with the Au heaters (Figure 5): fast and reproducible heating kinetics, excellent stretchability, and no noticeable deterioration over on/off cycling. However, some minor differences were observed between the two types of heaters. The temperature-time profile of composite heaters (Figure 5A) shows a slightly lower stable temperature and slower heating rate for each applied voltage. The temperature increases at a rate of 0.20 °C/s and 0.55 °C/s under 1.0 V and 1.7 V, respectively. This difference is attributed to the difference in heat transfer between the central Au resistive layer and the 5 nm ITO layer, which ultimately limits heat transfer to the thermal paste. The temperature versus strain profiles (Figure 5B,C) indicate that the multilayer thin film heaters have slightly less reproducibility as evidenced by the larger error bars obtained from replicate devices, especially at higher applied voltages. The lower reproducibility of the multilayer heaters at higher voltages could be caused by the two 5 nm ITO layers that are more brittle and could lead to some cracking during the stretching process.

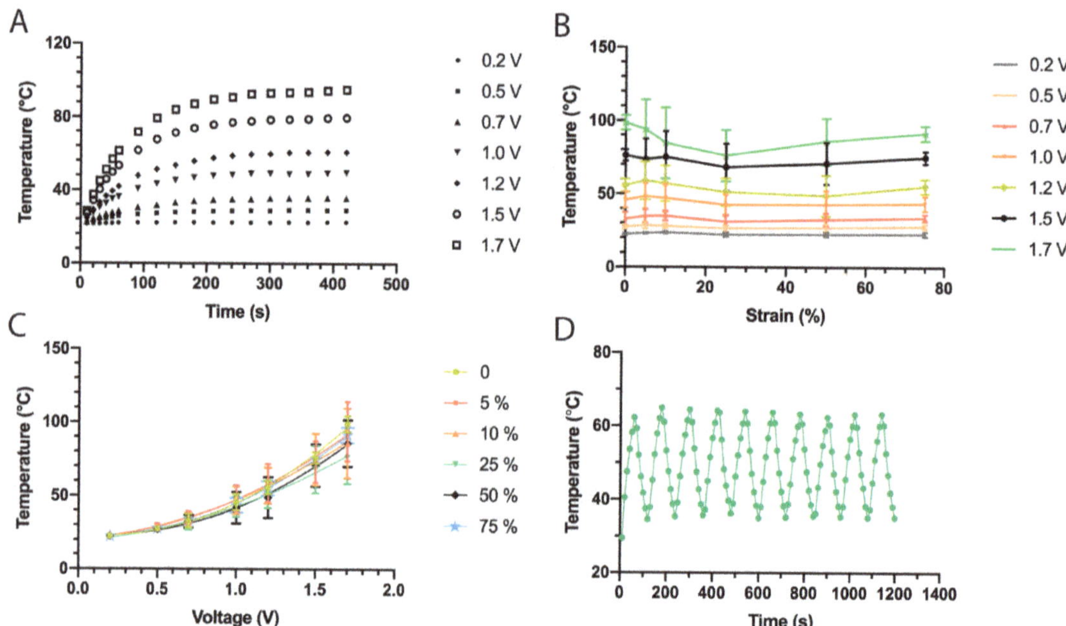

Figure 5. Characterization of stretchable ITO/Au/ITO heaters. (**A**) Temperature profiles under varying applied voltage (0.2–1.7 V). (**B**) The plateau temperature changes with various voltages but remains similar for any given voltage under different strains. (**C**) Changes in temperature vs. voltage for various applied strains (each series represents a given strain value). (**D**) Cycling on/off response under an applied voltage of 1.5 V shows the reproducibility of the temperature reached at the plateau.

It is worth mentioning that the ITO/Au/ITO multilayer heater is representative of an electronic device with a hybrid structure incorporating various materials (metal and ceramic); together with the contact pads which interconnect with other components, they comprise a unit cell that can be integrated into an array or a more complex circuit. This fabrication approach can be further extended to encompass other device components like resistors, capacitors, inductors, transistors, and even a hybrid network of them can be realized. The simultaneous transfer of multiple device components would also address interconnection issues that arise for stretchable electronic arrays, leading to improved long-term stability.

4. Conclusions

This work demonstrates a method for the fabrication of thin film stretchable devices by solvent-assisted transfer of wrinkled thin films from rigid substrates to elastomeric PDMS. Wrinkled thin films were obtained by shrinking flat films patterned on and supported by a shape memory polymer. The transfer process involves modifying the surface of the wrinkled film with an adhesion promoter that forms a chemical bond between PDMS and the film, and the subsequent release of the structured film through dissolution of a sacrificial layer or softening of the PS substrate. Structured films transferred using these two approaches were characterized and compared. The solvent-assisted transfer process ensures the faithful transfer of the wrinkled structures with millimeter to micrometer dimensions in the stretchable devices.

A proof-of-concept application was demonstrated by fabricating and characterizing stretchable thin film Au and ITO/Au/ITO multilayer stretchable heaters. The fabrication of highly stretchable heaters is simple, and since the connection pads and the heating elements were already overlayed, the transfer approach results in very strong adhesion between all elements of the heater. To measure thermal properties and stretchability of the heaters, the temperature of thermal paste in contact with the heaters was monitored at various applied strains and voltages. The results show that both types of stretchable thin films have rapid Joule heating kinetics, and the temperature can reach more than 50 °C with only 1 V of applied voltage. The heaters also displayed high stretchability with nearly constant performance even at 75% applied strain, and high reproducibility when subjected to on/off duty cycles. Overall, the transfer process is a simple and effective method for fabricating hybrid materials and structures for stretchable electronic devices.

Supplementary Materials: The following supporting information can be downloaded at: https://www.mdpi.com/article/10.3390/mi13020334/s1, Figure S1: Schematic of the characterization of stretchable Au heaters and ITO/Au/ITO heaters; Figure S2: Non-linear regression performed showing the quadratic fits presented in Figures 4C and 5C for stretchable Au heaters and ITO/Au/ITO heaters.

Author Contributions: X.D. designed and performed all experiments, analyzed the data, and wrote the first draft of the manuscript. J.M.M.-M. conceptualized the work, reviewed experimental data, edited the manuscript, and secured funding. All authors have read and agreed to the published version of the manuscript.

Funding: X.D. was partially supported through a China Scholarship Council doctoral award. J.M.M.-M. is the Tier 2 Canada Research Chair in Micro and Nanostructured Materials and the recipient of an Early Researcher Award from the Ontario Ministry of Research and Innovation. This work was supported by funding through a Discovery Grant from NSERC and a Canadian Foundation for Innovation J.E.L.F. instrumentation grant to J. M.-M. (RGPIN-2019-06433). This research made use of instrumentation within McMaster's Canadian Centre for Electron Microscopy.

Institutional Review Board Statement: Not Applicable.

Data Availability Statement: The data presented in this study are available on request from the corresponding author.

Acknowledgments: The authors thank Yujie Zhu for instrument training and Zhilin Peng for help with SEM imaging.

Conflicts of Interest: The authors declare no conflict of interest.

References

1. Rogers, J.A.; Someya, T.; Huang, Y. Materials and mechanics for stretchable electronics. *Science* **2010**, *327*, 1603–1607. [CrossRef] [PubMed]
2. Rogers, J.A.; Huang, Y. A curvy, stretchy future for electronics. *Proc. Natl. Acad. Sci. USA* **2009**, *106*, 10875–10876. [CrossRef] [PubMed]

3. Wagner, S.; Bauer, S. Materials for stretchable electronics. *MRS Bull.* **2012**, *37*, 207–213. [CrossRef]
4. Sekitani, T.; Someya, T. Stretchable organic integrated circuits for large-area electronic skin surfaces. *MRS Bull.* **2012**, *37*, 236–245. [CrossRef]
5. Kim, D.H.; Rogers, J.A. Stretchable electronics: Materials strategies and devices. *Adv. Mater.* **2008**, *20*, 4887–4892. [CrossRef]
6. Silvera-Tawil, D.; Rye, D.; Velonaki, M. Artificial skin and tactile sensing for socially interactive robots: A review. *Rob. Auton. Syst.* **2015**, *63*, 230–243. [CrossRef]
7. Suo, Z.; Ma, E.Y.; Gleskova, H.; Wagner, S. Mechanics of rollable and foldable film-on-foil electronics. *Appl. Phys. Lett.* **1999**, *74*, 1177–1179. [CrossRef]
8. Jiang, H.; Khang, D.-Y.; Song, J.; Sun, Y.; Huang, Y.; Rogers, J.A. Finite deformation mechanics in buckled thin films on compliant supports. *Proc. Natl. Acad. Sci. USA* **2007**, *104*, 15607–15612. [CrossRef] [PubMed]
9. Bowden, N.; Brittain, S.; Evans, A.G.; Hutchinson, J.W.; Whitesides, G.M. Spontaneous formation of ordered structures in thin films of metals supported on an elastomeric polymer. *Nature* **1998**, *393*, 146–149. [CrossRef]
10. Kim, D.H.; Ahn, J.-H.; Choi, W.M.; Kim, H.-S.; Kim, T.-H.; Song, J.; Huang, Y.Y.; Liu, Z.; Lu, C.; Rogers, J.A. Stretchable and foldable silicon integrated circuits. *Nature* **2008**, *320*, 507–511. [CrossRef] [PubMed]
11. Fu, C.C.; Grimes, A.; Long, M.; Ferri, C.G.L.; Rich, B.D.; Ghosh, S.; Ghosh, S.; Lee, P.; Gopinathan, A.; Khine, M. Tunable nanowrinkles on shape memory polymer sheets. *Adv. Mater.* **2009**, *21*, 4472–4476. [CrossRef]
12. Kim, J.; Park, S.-J.; Nguyen, T.; Chu, M.; Pegan, J.D.; Khine, M. Highly stretchable wrinkled gold thin film wires. *Appl. Phys. Lett.* **2016**, *108*, 061901. [CrossRef] [PubMed]
13. Zhu, Y.; Moran-Mirabal, J. Highly bendable and stretchable electrodes based on micro/nanostructured gold films for flexible sensors and electronics. *Adv. Electron. Mater.* **2016**, *2*, 1500345. [CrossRef]
14. Lacour, S.P.; Jones, J.; Wagner, S.; Li, T.; Suo, Z. Stretchable interconnects for elastic electronic surfaces. *Proc. IEEE* **2005**, *93*, 1459–1467. [CrossRef]
15. Choi, S.J.; Park, J.K.; Hyun, W.; Kim, J.; Kim, J.; Lee, Y.B.; Song, C.; Hwang, H.J.; Kim, J.H.; Hyeon, T.; et al. Stretchable heater using ligand-exchanged silver nanowire nanocomposite for wearable articular thermotherapy. *ACS Nano* **2015**, *9*, 6626–6633. [CrossRef] [PubMed]
16. Genzer, J.; Groenewold, J. Soft matter with hard skin: From skin wrinkles to templating and material characterization. *Soft Matter* **2006**, *2*, 310–323. [CrossRef] [PubMed]
17. Stimpson, T.C.; Osorio, D.A.; Cranston, E.D.; Moran-Mirabal, J.M. Direct comparison of three buckling-based methods to measure the elastic modulus of nanobiocomposite thin films. *ACS Appl. Mater. Interfaces* **2021**, *13*, 29187–29198. [CrossRef] [PubMed]
18. Lee, J.N.; Park, C.; Whitesides, G.M. Solvent compatibility of poly(dimethylsiloxane)-based microfluidic devices. *Anal. Chem.* **2003**, *75*, 6544–6554. [CrossRef] [PubMed]

 micromachines

Article

Non-Contact Optical Detection of Foreign Materials Adhered to Color Filter and Thin-Film Transistor

Fu-Ming Tzu [1,*], Shih-Hsien Hsu [2] and Jung-Shun Chen [3]

[1] Department of Marine Engineering, National Kaohsiung University of Science and Technology, Kaohsiung 80543, Taiwan
[2] Department of Electrical Engineering, Feng Chia University, Taichung 40802, Taiwan; shihhhsu@fcuoa.fcu.edu.tw
[3] Department of Industrial Technology Education, National Kaohsiung Normal University, Kaohsiung 80201, Taiwan; jschen@nknu.edu.tw
* Correspondence: fuming88@nkust.edu.tw

Abstract: This paper describes the non-contact optical detection of debris material that adheres to the substrates of color filters (CFs) and thin-film transistors (TFTs) by area charge-coupled devices (CCDs) and laser sensors. One of the optical detections is a side-view illumination by an area CCD that emits a coherency light to detect debris on the CF. In contrast to the height of the debris material, the image is acquired by transforming the geometric shape from a square to a circle. As a result, the side-view illumination from the area CCD identified the height of the debris adhered to the black matrix (BM) as well as the red, green, and blue of a CF with 95, 97, 98, and 99% accuracy compared to the golden sample. The uncertainty analysis was at 5% for the BM, 3% for the red, 2% for the green, and 1% for the blue. The other optical detection, a laser optical interception with a horizontal alignment, inspected the material foreign to the TFT. At the same time, laser sensors intercepted the debris on the TFT at a voltage of 3.5 V, which the five sets of laser optics make scanning the sample. Consequently, the scanning rate reached over 98% accuracy, and the uncertainty analysis was within 5%. Thus, both non-contact optical methods can detect debris at a 50 μm height or lower. The experiment presents a successful design for the efficient prevention of a valuable component malfunction.

Keywords: foreign material; laser sensor; area charge-coupled device; color filter; thin-film transistor

Citation: Tzu, F.-M.; Hsu, S.-H.; Chen, J.-S. Non-Contact Optical Detection of Foreign Materials Adhered to Color Filter and Thin-Film Transistor. *Micromachines* **2022**, *13*, 101. https://doi.org/10.3390/mi13010101

Academic Editor: Chengyuan Dong

Received: 9 December 2021
Accepted: 7 January 2022
Published: 8 January 2022

1. Introduction

A thin-film transistor liquid crystal display (TFT-LCD) comprises a sandwiched profile in which the central layer of a liquid crystal is between two substrates. The upper substrate combines the colorful layers of red, green, and blue in a color filter (CF). In contrast, the lower substrate is embedded with a microelectrode pixel with a data line and gate line across an electrical circuit, making up the thin-film transistor (TFT) [1]. Once a forward current flows through the electrode, the voltage of the pixel shifts the liquid crystal to switch the light in its optic direction. At the same time, a light penetrates the polarizer to filter out the random light and adjusts the brightness and darkness of the display.

A high-power exposure machine engraves the pattern of the CF in a gap from 100 to 300 μm with a deep ultraviolet light [2–4]. The array check tester inspects the electro pixel of the electric circuit in the gap 50 μm between the tester and TFT substrate [5]. The spin coater works with various photoresistors to inject the thin-film substrate into the narrow tolerance range, just several centimeters high—the chromaticity provides a vivid hue, colorful saturation, and level lightness [6,7].

The TFT-LCD requires large displays, and the uncertain issues in the factory's products create a huge challenge. Significantly, the valuable photomask, the critical modulator, and the precious inject coater can cause inevitable scratches due to foreign materials [2].

An abnormal process results in foreign material on the glass surface of the substrate; consequently, accidents can happen that damage the component. Such accidents affect the performance of the electronic switch and chromatic variation in the valuable component simultaneously and without warning. Thus, this is concerning to the thin-film maker.

The debris is caused by the fragmented glass, whose shapes are similar to the shiny parts of the micro and colorful photoresistors. Certain components loosen their parts, which then automatically adhere to the glass substrate, and the other foreign material is from the random texture of the bunny suit. The statistics indicate that the debris consists of 70% fragmented glass, 20% metal particles, and 10% other materials [2]. Thus, this is a well-known production problem.

Various literature reviews have proposed a measurement height from a diversified field. Groot (2017) [8] explained that the vertical resolution was miscalculated in the measurement height provided by an interference microscope operating with a 100 Hz, 1000 × 1000 pixel camera by the sinusoidal waveform, converting the intensity to a heightmap. Consequently, the repeatability was 0.072 nm. Musaoglu et al. (2019) [9] employed an atomic force microscope (AFM) to measure the height of copper oxide (CuOx). As a result, the accurate measurement reached a mean grain height of 11 nm. Kaplan (2021) [10] researched the Si photodiode and material characterization of TiO_2 and proposed a different technique to measure the height with an AFM. The result was 3 nm high, and the accuracy was significant. Kaplan also used X-ray diffraction to observe the polycrystalline structure. Thomas (2021) [11] investigated optical surface topography measurement methods by utilizing a scanning interferometer. The surface measurement based on optical imaging relied on an objective lens to collect diffracted rays. Moreover, Liu and He (2018) [12] developed a vehicle height system to use the transmitter and a receiver of laser detection to measure the vehicle's height.

The paper utilizes non-contact optical inspection of the debris material by an area photosensor and laser optics to detect foreign materials, as compared to the literature review. As a result, our topology is quick and economical for detection and accuracy. Other methods, such as an AFM and SEM, measure the height. An AFM uses unique tiny probes to detect a particular interaction between the probe and the sample surface and utilizes a piezoelectric ceramic scanner with a three-axis displacement to scan the surface of the sample. Thus, an AFM can measure the nanometer scale even at a height of several angstroms. Moreover, an AFM uses Van der Waals forces to present the surface characteristics of a sample. Another frequent method of height measurement is a scanning electric microscope (SEM), which utilizes an electron beam as a light source and the electromagnetic field as a lens to obtain the surface morphology of an object by collecting, sorting, and analyzing information generated by the interaction of electrons and samples.

The proposed method in this paper can measure the height at the micro-scale and is suitable for the thin-film transistor of a factory. The paper presents an original method of utilizing fast and accurate detection to meet the requirements of mass production in a factory. A non-contact optical inspection detects foreign material in the leading TFT-LCD foundries. A side-view illumination engages the photosensor on the top to inspect the CF in the dark field. The other intercepted debris is detected on the TFT by laser scanning. Each topology significantly remedies the loss of valuable components.

2. Principle of Detected Debris on the TFT-LCD

This study utilizes a non-contact optical method to detect debris adhered to the glass substrate. The side-view illumination installed on the edges of the glass emits the opposite in the dark field. The top of the photosensor receives the reflectance shade if any suspensive debris adheres to the panel. At that moment, the mapping of the grayscale calculates the height of the protrusion on the basis of transforming the area from square to circle. Equation (1) is as follows:

$$\sum_{i=1}^{n} X^2 \times k \cong \frac{\pi}{4} D^2 \tag{1}$$

where D is the circle's diameter/height, X represents the length of the effective pixel of the square based on the threshold grayscale, and k is a ratio value to approach both areas compared to the sample's known height. Figure 1 illustrates the structure of the debris detection using the side-view illumination emitting a coherence light to the suspension protrusion in the dark field. Figure 1A demonstrates the area CCDs installed on the top at a specific optical work distance. The illustration shows the six sets of CCDs, #1–2, #3–4, and #5–6, indicating a field of view that covers the inspected platform depending on the inspected size of the glass generation. The side-view illumination emits non-UV light to irradiate the suspension debris. The working distance (WD) is set for the optimum resolution of the pixel size of the color filter. The granite platform is a support to prevent the deformation of the mechanism. This is a robust construction to support the glass sample. When the area CCD acquires a reflected image, the acquired dada automatic program (ADAP) can judge the protrusion and send out a warning. The protrusion found most often is debris in the CF from non-uniform photoresistors, such as the red, blue, green, and black shown in the figure.

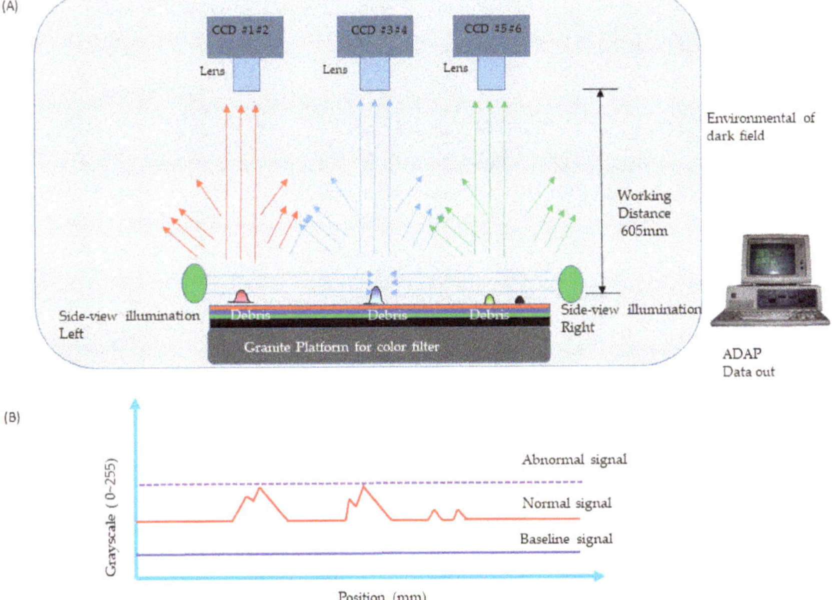

Figure 1. The struture of the photosensor that detects debris on the color filter (CF); (**A**) is a topology of the side-view illumination by the photosensor in the dark field, and (**B**) is the grayscale tendency.

Figure 1B demonstrates the grayscale tendency, indicating the base, normal, and abnormal signals during photosensor scanning. The abnormal signal expresses the height of the debris identified over the certainty height to compare the grayscale intensity. The topology applies to the CF; the background layer illustrates the non-transparent film. The experiment detects foreign material adhering to the substrate.

Furthermore, the surface of the glass guarantees that debris of no particular height adhere to the glass before the exposure machine; i.e., the gap is roughly 100~300 μm between the substrate and the exposure machine. When debris adheres to the CF, the ADAP detection completes the process to distribute the grayscale from 0 to 255. The granite platform maintains evenness to avoid variations due to changes in temperature. The photosensor takes the high-resolution 5 μm image (IMX204 CMOS color sensor, Sony Inc., Tokyo, Japan). The resolution is 5184 (H) × 3888 (V) at the number of recommended recording pixels. The magnification lens is 2×; thus, the optical resolution shifts to 10 μm, and, therefore, the detection is very powerful on the substrate glass. The structure detects foreign material quickly and accurately.

Moreover, the working distance (WD) expresses the height distance between the photosensor and objective for the optical resolution [12,13]. Equation (2) presents the thin-lens maker as follows:

$$WD = f \times \left(\frac{M+1}{M}\right) + f \times (M+1) + (\overline{HH^*}) \quad (2)$$

where f expresses the focal point to determine the image quality. The magnification of the image is presented by M; however, the coarse optical resolution is inverse to the image quality. The thickness of the optical lens in the photosensor is indicated by $\overline{HH^*}$. The photosensor converts the electrical signal to an image while scanning. A signal capacitor array triggers the photon electronic current in the photoactive region. The electrical current can amplify the signal to create an analog-to-digital conversion. Each capacitor accumulates electron charges in proportion to the light intensity. Alternatively, laser optics indicate foreign material adhered to the TFT in Equation (3):

$$u = \begin{cases} 3.5, & v \leq 3.5 \\ 5.0, & v = 5.0 \end{cases} . \quad (3)$$

Equation (3) demonstrates a voltage variation from 5 to 3.5 V once light travel is blocked, and the debris is adhered to the substrate. The illustration of the laser in Figure 2 presents two pieces of debris adhered to the TFT substrate. The task involves five sets of laser sensors with a coherency light to detect the debris from the transmitter to the receiver in Figure 2A. The illustration demonstrates the structure of laser optics that intercepts foreign debris and is moved by a servo motor to simultaneously drive the transmitter and receiver. The red line indicates the laser's beam, which travels to the inspected region. When the debris blocks the travel of the laser beam, the system detects the protrusion. The stage is robust enough to prevent deformation from external factors, such as temperature variations and stress and strain concentration. While the platform is moving, a sensor emits a spotlight from the transmitter. The platform is sixth-generation glass, 1800 mm × 1500 mm, which is a standard size in a TFT-LCD factory. Thus, the laser detects a sample and intercepts the debris by moving back and forth. The ADAP sends out a warning if the voltage is below 3.5 V.

In contrast to the CF, the TFT needs the laser to intercept the debris because the transparent film on the bottom layer reflects the background and interferes with the results. Thus, we chose a laser sensor to detect the debris, rather than the area CCD.

Furthermore, the light source of the laser is the red semiconductor at 670 nm (Model LX2-110W, Keyence Inc., Osaka, Japan). The transmitter emits the coherency light at 1~2.5 mm diameter of the spotlight to detect at a distance of 300~2000 mm. The response time is 0.5 ms, and the power is 12~24 VDC ± 10% ripple. Figure 2B indicates the profile of voltage variation. Once a voltage drop below 3.5 V is detected, a certain height is identified over the gap. The laser is an optical oscillator that generates light with coherence, directivity, and narrow width.

The standard deviation is known as the mean square deviation, and the mathematical symbol σ (sigma) is most commonly used in probability statistics to measure the degree of dispersion of a set of values. Thus, the higher the standard deviation, the more significant the difference among the data. This paper takes Equation (4) to analyze the difference using the ten times measurement, which quantifies the accuracy:

$$\sigma = \sqrt{\frac{1}{n-1}\sum_{1}^{n}(p_{i-}p)^2} \quad (4)$$

where $\overline{p}$ expresses a mean value of the data set, and the symbol n expresses the samples. The gauge of repeatability and reproducibility (R&R) shows the measurement's deviation [14–16]. The purpose of gauging R&R is to evaluate the accuracy of the measurement system and the personnel operation. Repeatability is a feature of one part measured multiple times to analyze the sum of each difference by the same operator. Reproducibility is a feature of the same part measured using measuring tools that analyze the sum of different operators [17,18]. Experimentally, the non-contact optical detection adopts the repeatability measurement [19,20].

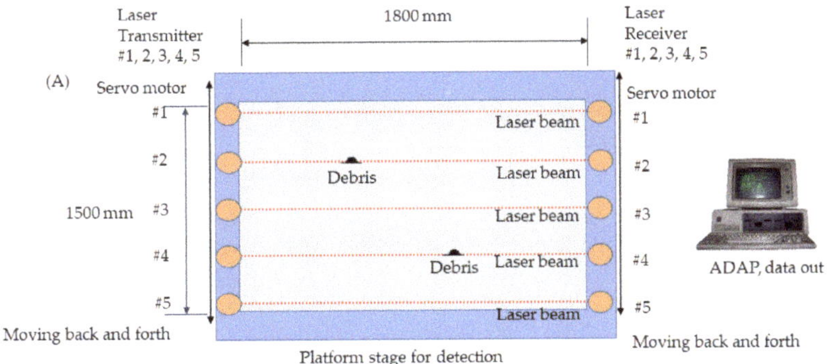

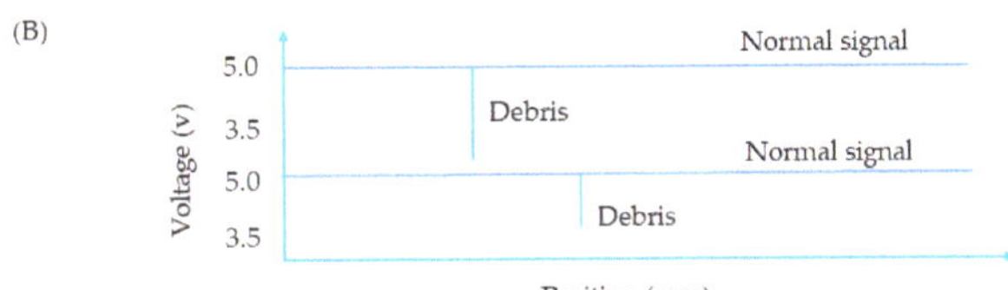

Figure 2. The architecture of the laser sensor illustrating the topology of a thin-film transistor (TFT), (**A**) is a topology of the foreign materials on TFT by laser optics (**B**) is a voltage tendency.

3. Experimental Architecture

The experiment used a photo image to calculate the height of the debris on the CF and performed laser scanning to intercept foreign material on the TFT in the clean Room 10 of the factory. Figure 3A illustrates the structure for detecting the foreign material on the CF glass. First, the glass was aligned before detection in (1). The station detected the debris to calculate the height of the protrusion with a photosensor, where the photosensor was installed on the top at (2). After completing the detection, the glass was delivered to Station (3) for UV light exposure, and then the glass was delivered to Station (4). Figure 3B illustrates an interceptive topology of the laser sensor. The glass was delivered to Station (1) for alignment. Station (2) was a scanning region for the laser sensor. Station (3) performed an array check with the electrical field applied modulator. This was a non-contact detection by rotating the liquid crystal in the electronic circuit of the electrode pixel that caused the light to switch on and off. Finally, the glass was delivered out of Station (4). The experiment's optical parameters are tabulated in Table 1. The parameter for the non contact optical detection of the debris on the CF indicated the CCD pixel at 5 μm optical resolution applied on sixth-generation glass. The substrate size was 1850 mm × 1500 mm, which enabled the area photosensor room to cover the detectible region for six sets of CCDs. The working distance was 605 mm.

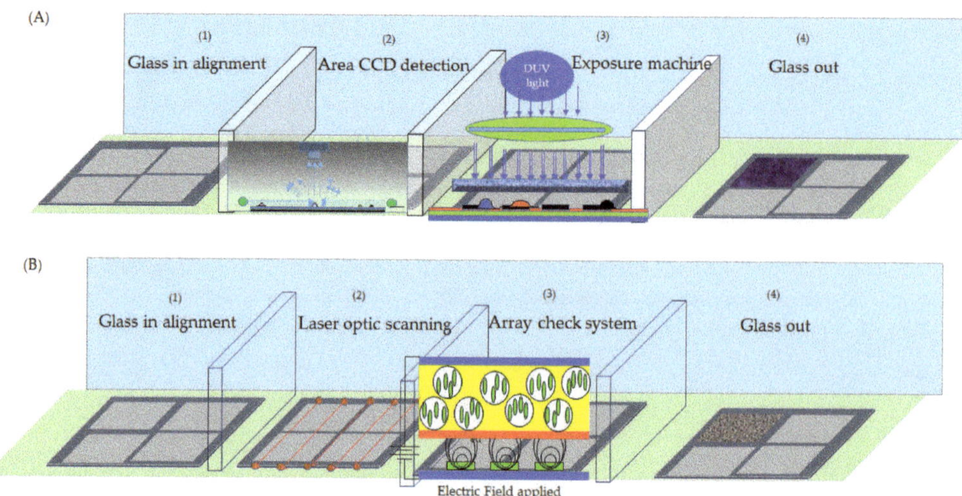

Figure 3. The architecture of the optical detection system comprising (**A**) of CF at (1) glass-in, (2) CCD detection, (3) exposure machine, (4) glass-out; below diagram (**B**) of TFT includes (1) glass-in, (2) laser optic scanning, (3) array check system, and (4) glass-out.

Table 1. The parameters for optical detection of the debris on the color filter.

Description	Value	Detailed	Length	Width
CCD Pixel Size (μm)	5	Substrate dimension (mm)	1850	1500
Magnification	2	Signal CCD FOV (mm)	1037	778
Optical resolution (μm)	10	Require CCD quantity	3	2
Focused lens (mm)	50	Signal overlap (mm)	252	14
Working distance (mm)	605	All overlap (mm)	533	750

Figure 4 illustrates the array check system for detecting an electrode pixel in the narrow gap of 50 μm between the modulator and the substrate. Significantly, the second-to-last layer on the modulator is made of the multi-layer pellicle mirror, which is mostly a combination of zirconium (Zr) and oxygen (O). The paper will discuss the details later.

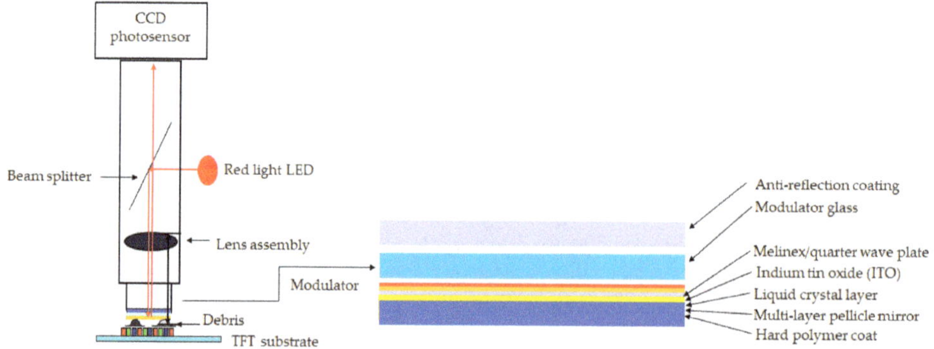

Figure 4. The topology of the array check system indicating the components, especially the modulator.

4. Result and Discussion

The experiment used the image of the CF taken by an area CCD and a laser sensor to detect the TFT debris from a factory. The task detected the height of the debris by engaging a side-view illumination to compare it with the known height from the golden sample. Figure 5 illustrates that the known height from the golden sample was 130 μm for the black matrix (BM), 131 μm for the red, 155 μm for the green, and 146 μm for the blue. Figure 5A indicates a 10× magnification of the raw image by microscope (MODEL: M-10X, Newport Inc., Irvine, CA, USA). The golden sample comprised the lead-free solder ball (S010, Unano Technology, Tainan, Taiwan). Figure 5B shows the image made by the photosensor. Figure 5C is the image enlarged 1600% by Photoshop software. Figure 6 indicates a grayscale intensity label in the color pixel of the image. As a result, the height of detection was 124 μm for the BM, 127 μm for the red, 151 μm for the green, and 145 μm for the blue. The uncertainty analysis was based on repeatability to acquire 10 times the data. Table 2 tabulates the repeatability of the BM, red, green, and blue measurements at 10 times. The measurement averages were 131, 135, 161, and 153 for the BM, red, green, and blue, respectively. After comparing the known heights from the golden sample, the result showed error levels of 5% for the BM, 3% for the red, 2% for the green, and 1% for the blue.

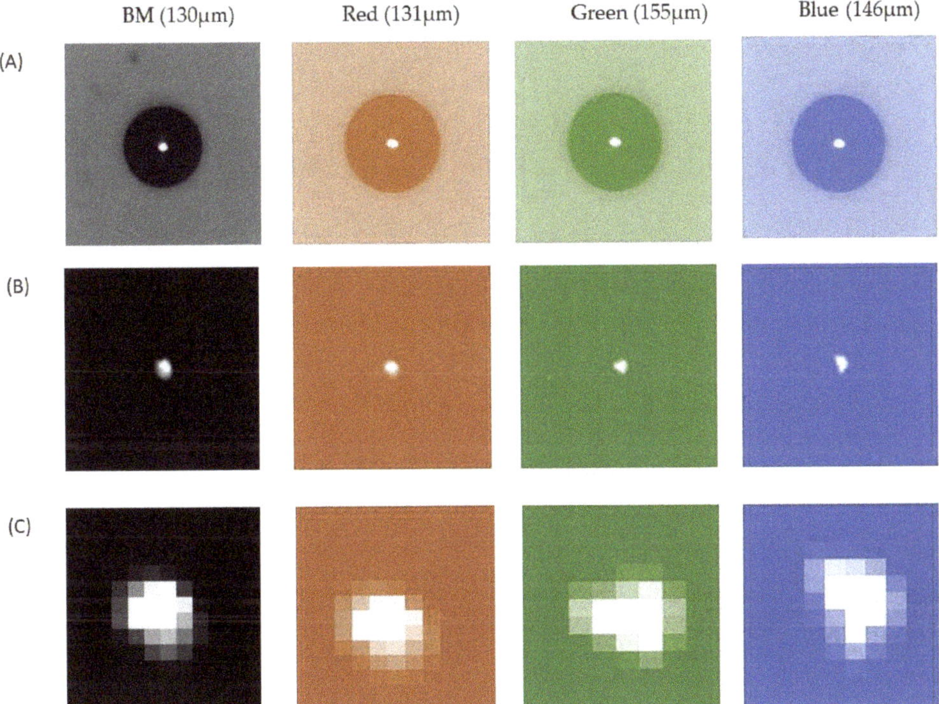

Figure 5. The golden sample coating the colored photoresists (PRs) with heights of 130, 131, 155, and 146 μm. Row (**A**) indicates the various solder balls magnified by microscope 10×, (**B**) indicates the grayscale image, and (**C**) is enlarged 1600% by Photoshop software.

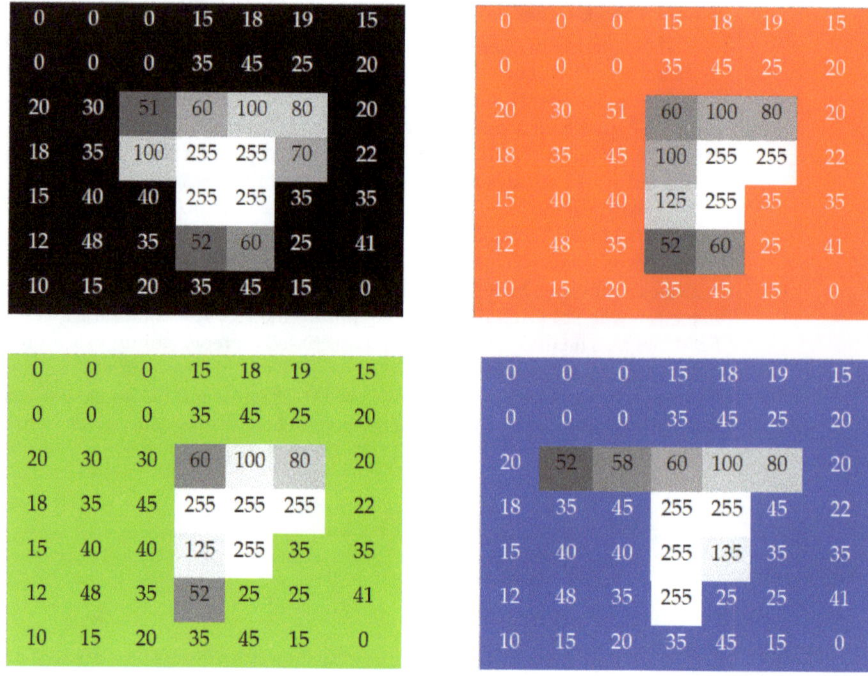

Figure 6. Images showing the grayscale with the threshold value for the black matrix (BM) as well as the red, green, and blue for the color filter (CF).

Table 2. The height measurement as indicated ten times compared to the golden sample, including uncertainty analysis.

Glass	1	2	3	4	5	6	7	8	9	10	Avg	Calculated (μm)	Known (μm)	Error	3σ
BM	125	132	130	132	130	132	133	130	132	133	131	124	130	5%	5%
Red	135	135	132	134	135	136	135	135	136	135	135	127	131	3%	3%
Green	160	160	162	163	160	160	162	160	160	160	161	152	155	2%	2%
Blue	153	153	155	153	153	153	153	153	153	154	153	145	146	1%	1%

Since the BM is a black layer that absorbs reflective light from the environment, the PR was shown as the weakest response. Thus, the error was higher than those for the colored PRs. The most common standard deviation was applied to the probability statistics to measure the degree of statistical dispersion. Statistically, the 68–95–99.7 rule was within the normal distribution [21]. The mean percentage concerned one standard deviation, two standard deviations, and three standard deviations. At the same time, the experiment took the three standard deviations (3σ) at a 99.7% confidence level [22,23].

Furthermore, the blue PR indicated that the minor deviation in the measurement was due to short wavelength with a significant reflective response. Consequently, the 3σ indicated this at 5% for the BM, 3% for the red, 2% for the green, and 1% for the blue.

The task filtered out UV wavelengths smaller than 400 nm to avoid overexposure of the thin film and to prevent the characteristic PR variance [24], as seen in Figure 7. The wavelength distribution illustrates the profiles for the BM, red, green, and blue. In addition, the air was a reference spectrum against which each PR response was compared. The BM demonstrated the weakest light response because of the black body. The red PR presented 40% intensity. The green PR demonstrated the highest intensity at over 60% response intensity because the green wavelength had the most vision sensitivity among the colors. The response of the blue light was around 40%.

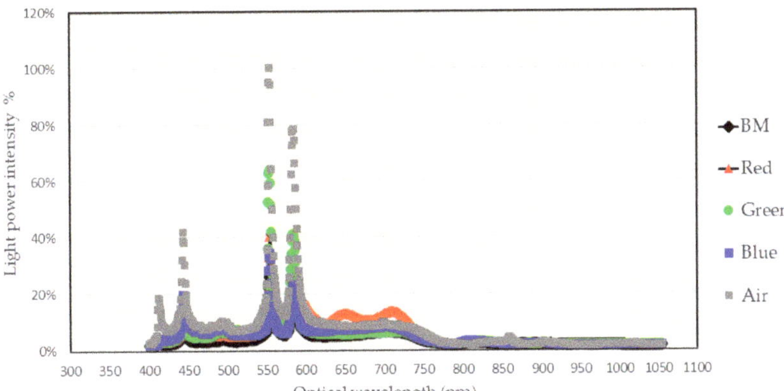

Figure 7. The distribution of the spectrum measures of the photoresists for the black matrix (BM), red, green, and blue.

Figure 8 illustrates the uniform light emitted from the region that connects the multiple fibers to the light source. The start-and-end position tended toward a sloop shape; thus, we chose the uniformity brightness as a detection region.

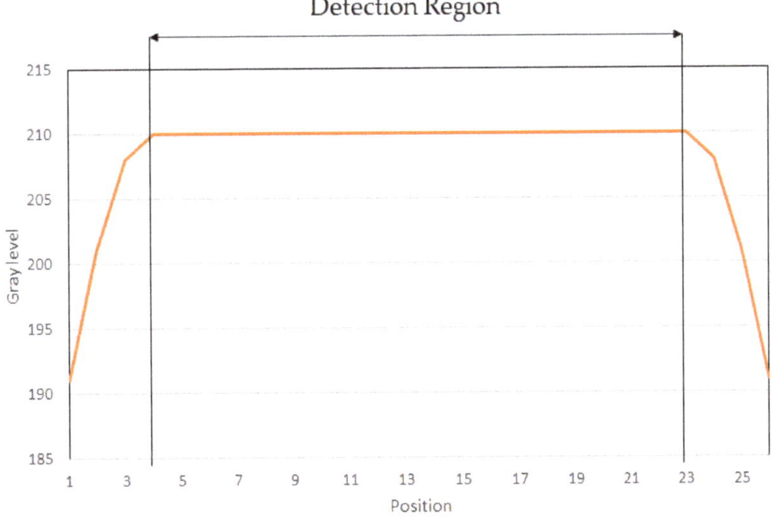

Figure 8. The uniformity of the side-view illumination indicating the distribution.

The experiment used the laser sensor to scan the whole region. Figure 9 shows the laser's tendency to intercept the debris. The horizontal lines express the scanning length using five sets of coherence light-emitting sensors across the receiver. We specifically installed a granite platform with optimized flatness to avoid the warpage. The vertical lines indicate voltage variation. Once the voltage was below 3.5 V, it was considered a successful interception on the glass of the substrate.

Moreover, the laser beam was a coherence light that traveled the detected region from transmitter to receiver above the 100 µm height of the substrate. It was an excellent choice to apply the transparent and non-transparent thin films. Otherwise, we had not used the laser optics, the reflective image with a sub-layer pattern of the TFT would have interfered with detection. Conversely, we developed the optical detection to apply laser optics to the thin-film transistors.

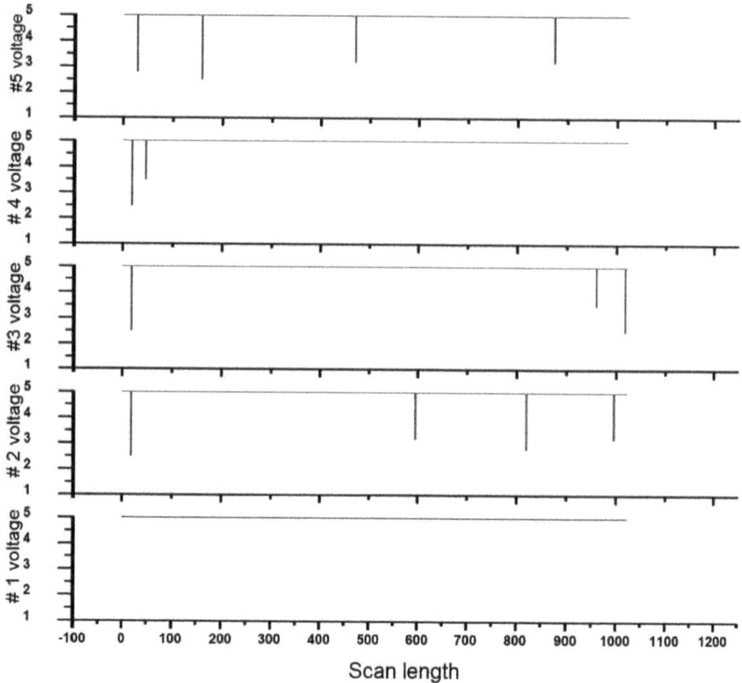

Figure 9. The profile of the laser sensor indicating the five sets of receivers.

Moreover, the voltage for sensor #1 indicated the normal condition at 5 V. Sensors #2–5 intercepted the debris at 4, 3, 2, and 4. Accordingly, the voltage was below 3.5 V. The experiment was undertaken ten times for accuracy of the statistics. The measurements are tabulated in Table 3. The height of the golden sample is in the table. Consequently, the maximum 5% for the 3σ was a 99.7% confidence interval.

Table 3. Statistics of the uncertainty indicating the ten measurements for three standard deviations (3σ).

Laser #	1	2	3	4	5	6	7	8	9	10	Average	Height (um)	3σ (%)
1	4.99	4.99	4.95	4.99	4.95	4.99	4.95	4.90	4.98	4.98	4.97	100	2%
2	2.55	2.45	2.45	2.52	2.55	2.44	2.55	2.50	2.45	2.45	2.49	200	5%
2	3.23	3.25	3.15	3.25	3.24	3.33	3.22	3.28	3.26	3.25	3.25	200	4%
2	2.81	2.85	2.88	2.85	2.75	2.81	2.80	2.80	2.85	2.75	2.82	200	4%
2	3.22	3.15	3.23	3.25	3.15	3.15	3.25	3.21	3.10	3.15	3.19	200	5%
3	2.54	2.55	2.56	2.45	2.47	2.48	2.48	2.49	2.45	2.45	2.49	250	5%
3	3.48	3.52	3.55	3.45	3.45	3.56	3.47	3.41	3.42	3.55	3.49	250	5%
3	2.51	2.52	2.52	2.55	2.53	2.55	2.55	2.45	2.52	2.51	2.52	250	3%
4	2.45	2.45	2.55	2.45	2.55	2.45	2.45	2.50	2.45	2.44	2.47	260	5%
4	3.52	3.45	3.55	3.45	3.55	3.42	3.52	3.41	3.46	3.55	3.49	260	5%
5	2.82	2.78	2.77	2.81	2.88	2.85	2.85	2.82	2.88	2.88	2.83	280	4%
5	2.52	2.55	2.45	2.45	2.52	2.51	2.55	2.52	2.47	2.48	2.50	280	4%
5	3.25	3.22	3.32	3.27	3.28	3.25	3.26	3.25	3.22	3.21	3.25	280	3%
5	3.26	3.25	3.15	3.25	3.22	3.24	3.25	3.21	3.29	3.28	3.24	280	3%

Furthermore, the modulator of the array check system was a critical component for the liquid crystal layer to inspect the electrode pixel at the dimensions of 143 mm × 133 mm. An energy-dispersive X-ray spectroscopy (EDX) was used to detect the elemental analysis and chemical characterization. The experiment was conducted with a field emission scanning electron microscope (FESEM) manufactured by PhotoMetrics, Inc. (Huntington Beach, CA, USA), which analyzed the spectrum for the modulator. Since elements emit different emission spectra because of their different atomic structures, the components were distinguished by their X-ray spectra. As a result, the most significant elements on the layer of the modulator were oxygen (O) and zirconium (Zr), i.e., O35.51Zr32.64B14.42 Si8.38C9.05 (wt.%), as investigated by EDX.

Zr is a metal element used mainly for heat-resistance and as a sunscreen agent, while a small amount of zirconium is an alloying agent to resist corrosion. Zr is the prevalent metal in electronic instruments. A Zr foil was applied in a thin film, including chemical vapor deposition (CVD) and physical vapor deposition (PVD). The thickness of the foil was approximately 2 mm on the modulator. A hard polymer coat was the bottom layer, and a multi-layer pellicle of the Zr was the second-to-last layer. Figure 10 shows the EDX spectrum, and Table 4 tabulates the presence of the constituent elements statistically.

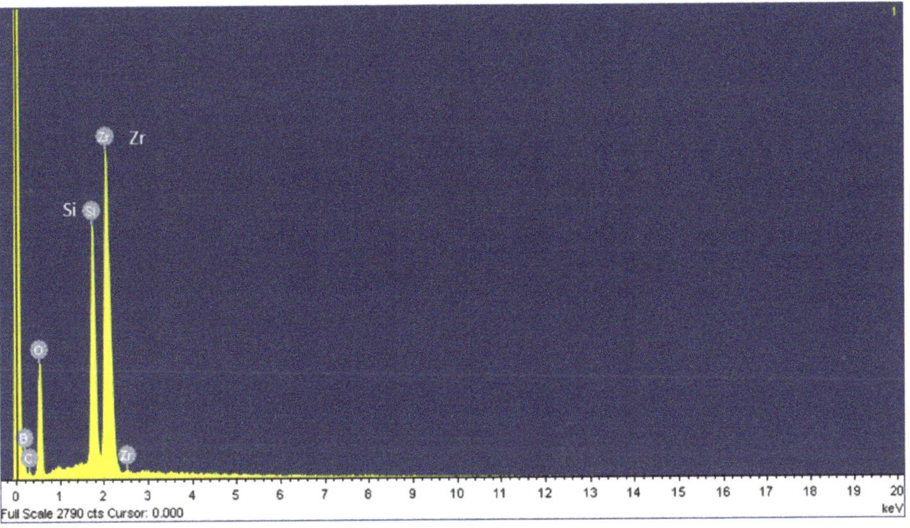

Figure 10. The EDX spectrum of the FISEM indicating the response of the material.

Table 4. Statistics of the element combinations tabulating the weight %.

Element	O	Zr	B	Si	C	Total
Mean	35.51	32.64	14.42	8.38	9.05	100
Max.	35.51	32.64	14.42	8.38	9.05	100
Min.	35.51	32.64	14.42	8.38	9.05	100
Std. deviation	0	0	0	0	0	0

In particular, the experiment investigated the damaged condition of the array check system modulator with a photosensor. Figure 11 illustrates the (A) typical structure of the modulator, (B) indicates a scratch across the area and several debris pieces adhered to the surface, and (C) is an enlarged view.

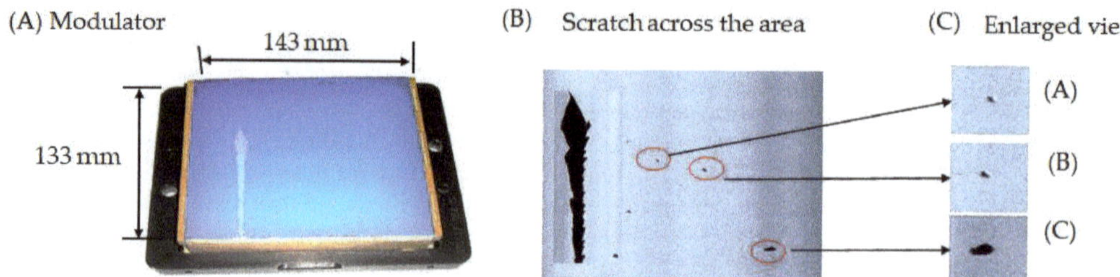

Figure 11. Analysis of the debris adhering to the modulator indicates (**A**) of the damaged modulator, (**B**) of scratch across the area, and (**C**) of the enlarged view.

Figure 12 shows enlarged views of the raw and grayscale images to investigate the debris on the modulator. The images underwent a magnification of 400% to show the grayscale distribution. As a result, the calculated heights (A) 123 μm, (B) 245 μm, and (C) 687 μm were the bases of a threshold of 50 for the golden sample by transforming the square to the circle. Table 5 indicates the known heights (A) 125 μm, (B) 250 μm, and (C) 700 μm. Thus, the calculated height exceeded the known height by an error of 2% with the 3σ of 2% for the repeatability measurement. Thus, the non-contact optical detection provided high accuracy, and the damage-free significance was very impressive.

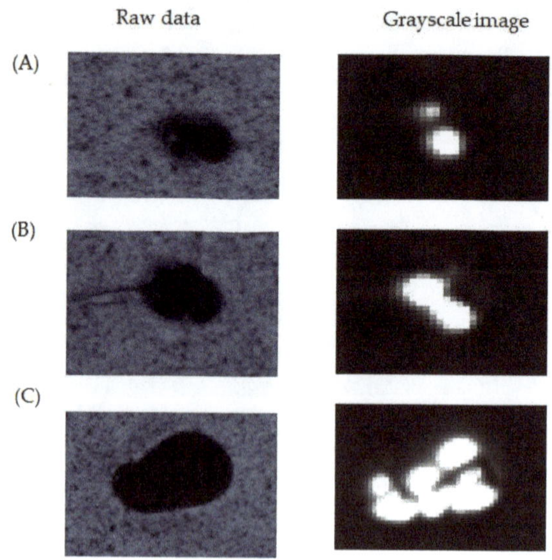

Figure 12. Images of the debris enlarged by the microscope indicate the known heights (**A**) 125 μm, (**B**) 250 μm, and (**C**) 700 μm.

Table 5. Measurements of the debris on the modulator.

Debris	1	2	3	4	5	6	7	8	9	10	Avg	Calculated (um)	Known (um)	Error	3σ
A	132	130	130	130	130	129	130	128	130	129	130	123	125	2%	2%
B	260	259	260	258	260	260	258	260	256	260	259	245	250	2%	2%
C	725	730	730	735	730	730	720	725	725	730	728	687	700	2%	2%

5. Conclusions

This study experimentally investigated the non-contact optical method for detecting the height of debris on CF glass by engaging a side-view illumination using the area CCD and a laser sensor emitting a coherency light to scan the TFT. The result showed the CF accuracy was at 5% for the BM, 3% for red, 2% for green, and 1% for blue. In addition, the repeatability of the three standard deviations was at 5% for the BM, 3% for the red, 2% for the green, and 1% for the blue. Alternatively, the laser intercepting the certain height of debris adhered to the TFT was indicated at a voltage below 3.5 V. The accuracy of the repeatability was also within 5%.

Moreover, the calculated debris height indicated a 2% error over the known height and a 2% repeatability at the three standard deviations on the modulator. Thus, an optical inspection can ensure the damage-free condition of the critical component. In other words, the non-contact optical inspection extends the necessary applications.

Author Contributions: Data curation, F.-M.T. and J.-S.C.; formal analysis, F.-M.T. and S.-H.H.; methodology, F.-M.T., J.-S.C., and S.-H.H.; validation, F.-M.T.; writing—original draft, F.-M.T.; writing—review and editing, F.-M.T. All authors have read and agreed to the published version of the manuscript.

Funding: This research received no external funding.

Institutional Review Board Statement: Not applicable.

Informed Consent Statement: Not applicable.

Data Availability Statement: The data presented in this study are available on request from the corresponding author.

Conflicts of Interest: The authors declare no conflict of interest.

References

1. Jiang, H.H.; Xiao, J.; Yin, Z.Y.; Zhang, L.; Yang, H.F.; Gao, X.; Wang, S.D. Progress and Outlook on Electron Injection in Inverted Organic Light-Emitting Diodes. *Chin. Sci. Bull.-Chin.* **2021**, *66*, 2105–2116. [CrossRef]
2. Tzu, F.M.; Chen, J.S.; Chou, J.H. Optical Detection of Protrusive Defects on a Thin-Film Transistor. *Crystals* **2018**, *8*, 440. [CrossRef]
3. Koppelhuber, A.; Bimber, O. Multi-Exposure Color Imaging with Stacked Thin-Film Luminescent Concentrators. *Opt. Express* **2015**, *23*, 33713–33720. [CrossRef] [PubMed]
4. Yen, T.C.; Tso, P.L. Fine Line-Width Black Matrix of a Color Filter by an Advanced Polishing Method. *J. Micromech. Microeng.* **2004**, *14*, 867–875. [CrossRef]
5. Wang, Y.C.; Lin, B.S.; Chan, M.C. Electro-Optical Measurement and Process Inspection for Integrated Gate Driver Circuit on Thin-Film-Transistor Array Panels. *Measurement* **2015**, *70*, 83–87. [CrossRef]
6. Seo, M.; Kim, J.; Oh, H.; Kim, M.; Baek, I.U.; Choi, K.D.; Byun, J.Y.; Lee, M. Printing of Highly Vivid Structural Colors on Metal Substrates with a Metal-Dielectric Double Layer. *Adv. Opt. Mater.* **2019**, *7*, 1900196. [CrossRef]
7. Wang, Y.S.; Zheng, M.J.; Ruan, Q.F.; Zhou, Y.M.; Chen, Y.Q.; Dai, P.; Yang, Z.M.; Lin, Z.H.; Long, Y.; Li, Y.; et al. Stepwise-Nanocavity-Assisted Transmissive Color Filter Array Microprints. *Research* **2018**, *2018*, 8109054. [CrossRef] [PubMed]
8. de Groot, P.J. The Meaning and Measure of Vertical Resolution in Optical Surface Topography Measurement. *Appl. Sci.* **2017**, *7*, 54. [CrossRef]
9. Musaoglu, C.; Pat, S.; Mohammadigharehbagh, R.; Ozen, S.; Korkmaz, S. The Thermionic Vacuum Arc Method for Rapid Deposition of Cu/Cuo/Cu2o Thin Film. *J. Electron. Mater.* **2019**, *48*, 2272–2277. [CrossRef]
10. Kaplan, H.K.; Olkun, A.; Akay, S.K.; Pat, S. Si-Based Photodiode and Material Characterization of TiO_2 Thin Film. *Opt. Quantum Electron.* **2021**, *53*, 248. [CrossRef]
11. Thomas, M.; Su, R.; de Groot, P.; Coupland, J.; Leach, R. Surface Measuring Coherence Scanning Interferometry Beyond the Specular Reflection Limit. *Opt. Express* **2021**, *29*, 36121–36131. [CrossRef]
12. Liu, X.; He, R. Application of Laser Sensor in Urban Vehicle Type Detection. *Teh. Vjesn.-Tech. Gaz.* **2018**, *25*, 622–626.
13. La Spada, L.; Vegni, L. Electromagnetic Nanoparticles for Sensing and Medical Diagnostic Applications. *Materials* **2018**, *11*, 603. [CrossRef] [PubMed]
14. Osthus, D.; Weaver, B.P.; Casleton, F.M.; Hamada, M.S.; Steiner, S.H. On Gauge Repeatability and Reproducibility Studies for Counts. *Qual. Eng.* **2021**, *33*, 641–654. [CrossRef]
15. Schaltegger, U.; Ovtcharova, M.; Gaynor, S.P.; Schoene, B.; Wotzlaw, J.F.; Davies, J.; Farina, F.; Greber, N.D.; Szymanowski, D.; Chelle-Michou, C. Long-Term Repeatability and Interlaboratory Reproducibility of High-Precision Id-Tims U-Pb Geochronology. *J. Anal. At. Spectrom.* **2021**, *36*, 1466–1477. [CrossRef]

16. Snee, R.D. Interpreting Operator-Part Interaction in Gage Repeatability and Reproducibility Studies. *Qual. Eng.* **2021**, *33*, 395–403. [CrossRef]
17. Huang, L.; Wang, T.Y.; Nicolas, J.; Polack, F.; Zuo, C.; Nakhoda, K.; Idir, M. Multi-Pitch Self-Calibration Measurement Using a Nano-Accuracy Surface Profiler for X-Ray Mirror Metrology. *Opt. Express* **2020**, *28*, 23060–23074. [CrossRef]
18. Nunes, M.C.S.; Lima, L.S.; Yoshimura, E.M.; Franca, L.V.S.; Baffa, O.; Jacobsohn, L.G.; Malthez, A.; Kunzel, R.; Trindade, N.M. Characterization of the Optically Stimulated Luminescence (Osl) Response of Beta-Irradiated Alexandrite-Polymer Composites. *J. Lumin.* **2020**, *226*, 117479. [CrossRef]
19. Pues, P.; Schwung, S.; Rytz, D.; Justel, T. On the Use of Luminescent Single Crystals as Optical Reference Materials. *J. Lumin.* **2021**, *238*, 118289. [CrossRef]
20. Wang, C.R.; Chen, M.Q.; Zhao, Y.; He, T.Y.; Liu, Q.F. Low Temperature Crosstalk Optical Fiber Fabry-Perot Interferometers for Highly Sensitivity Strain Measurement Based on Parallel Vernier Effect. *Opt. Fiber Technol.* **2021**, *67*, 102700. [CrossRef]
21. Wang, T.M.; Shih, Z.C. Measurement and Analysis of Depth Resolution Using Active Stereo Cameras. *IEEE Sens. J.* **2021**, *21*, 9218–9230. [CrossRef]
22. Ding, C.S.; Tamma, K.K.; Lian, H.J.; Ding, Y.J.; Dodwell, T.J.; Bordas, S.P.A. Uncertainty Quantification of Spatially Uncorrelated Loads with a Reduced-Order Stochastic Isogeometric Method. *Comput. Mech.* **2021**, *67*, 1255–1271. [CrossRef]
23. Levental, M.Y.; Pogodin, Y.M.; Mironov, Y.R. Improving the Methodology for Calculation of Energy Losses in Axial Turbine Cascades. *Mar. Intellect. Technol.* **2021**, *2*, 104–109.
24. Augustin, A.J. Reliable Uv-Light Protection in Intraocular Lenses—Scientific Rationale and Quality Requirements. *Klin. Mon. Fur Augenheilkd.* **2014**, *231*, 901–908.

www.ingramcontent.com/pod-product-compliance
Lightning Source LLC
LaVergne TN
LVHW070659170726
843515LV00004B/451

* 9 7 8 3 0 3 6 5 5 2 9 4 1 *

MDPI
St. Alban-Anlage 66
4052 Basel
Switzerland
Tel. +41 61 683 77 34
Fax +41 61 302 89 18
www.mdpi.com

Micromachines Editorial Office
E-mail: micromachines@mdpi.com
www.mdpi.com/journal/micromachines